PILOT'S
HANDBOOK of WEATHER

PILOT'S

HANDBOOK of WEATHER

by
Lt. Col. Gene Guerny
and
Capt. Joseph A. Skiera

SECOND EDITION
Revised by
L. W. Reithmaier

1974

AERO PUBLISHERS, INC.

329 Aviation Road   Fallbrook, Cal. 92028

Library of Congress Catalog Card Number
74-77535

ISBN 0-8168-7355-0

Printed and Published in the United States by Aero Publishers, Inc.

PREFACE TO SECOND EDITION

Although the basics of weather have changed little over the years, important advances have been made in the study of meteorology, gathering and processing of data, forecasting procedures and the presentation of weather information to the pilot.

These advances are reflected in this second edition of *Pilot's Handbook of Weather*. This volume thus provides an up-to-date and expanded text for pilots and others whose interest in meteorology is primarily in its application to flying.

Every pilot needs to know something about the behavior of weather and how various weather conditions affect flying. Just as he does not need to be an aeronautical engineer to fly, neither does he need to become a fully trained meteorologist to cope with weather. The pilot, however, should have a practical understanding of those meteorological principles important to aviation. This is essential to his effective use of current and forecast weather information. Above all, the pilot needs to know how to use his total weather knowledge to best advantage in flying safely and efficiently.

This second edition of *Pilot's Handbook of Weather* provides the pilot with the necessary weather background, how to evaluate weather data and how to effectively cope with weather in flight.

ACKNOWLEDGEMENT

Greatest appreciation is voiced for the weather materials supplied from the Department of The Air Force's manual "Weather for Aircrews" and the U. S. NAVY'S Chief of Naval operations' publication "All-Weather Flight Manual".

Gene Guerny
Joe Skiera

For the second edition, additional material was utilized from various Federal Aviation Administration (FAA) publications, the National Weather Service and the RCA Aviation Equipment Department.

L. W. Reithmaier

PHOTO CREDIT

TABLE OF CONTENTS

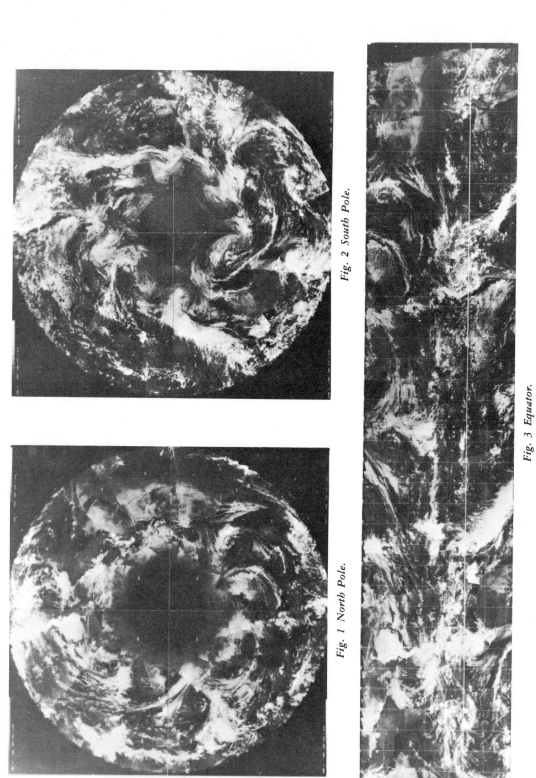

Fig. 2 South Pole.

Fig. 3 Equator.

Fig. 1 North Pole.

Sattelite computerized mosaic of the world's weather for one day.

Fig. 4 Weather briefing is a required part of pre-flight planning.

Weather cannot be read on a dial. Weather elements such as cloud form, temperature, and turbulence have the same relation to weather as instrument readings have to an airplane's equipment. Any pilot who does not know how to use the information gained from these weather elements is in about the same predicament as he would be if he did not know the meaning of the information given by the instruments on his airplane's panel.

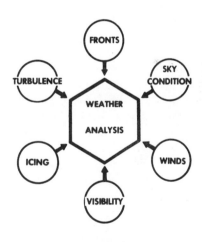

Fig. 6 Weather analysis.

Fig. 5 Aircraft and instruments.

Proper understanding and use of weather information is the responsibility of every pilot.

The weather forecaster is at the service of the pilot for the purpose of aiding in flight planning, but it is the pilot's responsibility to make full use of available information and to know his weather.

The atmosphere is the airman's medium. To become a good airman, a pilot must acquire the habit of being constantly alert to the weather around him. He must reach the point where, on the ground or in the air, he subconsciously notes changes in weather elements, such as wind, temperature, and clouds, just as he is constantly aware of what city he is in and of the time of day.

Specific duties of the pilot with respect to weather may be stated as follows:

1. Before flight he must get a picture of the weather that is expected over the whole area that may be covered during the proposed

flight. This does not mean just the weather along one line at a certain altitude, but the weather over the whole area at all possible operating altitudes. He should decide before the flight what to do in case unexpected trouble is encountered.

2. During flight, the forecast of flight weather must be constantly checked against observations. Such a check will usually verify the correctness of the forecast. Occasionally, the check will show that the forecast has been "busted" — gone wrong.

3. When a forecast has gone wrong, a decision, based on an understanding of the weather picture gained before take-off and on an interpretation of the weather observed during the flight, must be made. No general rules can be given to solve the various problems likely to arise.

4. Accurate and intelligent reports of the weather encountered should be kept. Because of the limitations imposed upon the distribution of ground weather reporting stations, weather reports brought back by pilots operating over areas from which other reports are not available are of great value to the forecaster and to other pilots flying in the same area.

"VFR" AND "IFR"

Most pilots fly VFR or under Visual Flight Rules. Weather conditions for VFR flight are defined in the Federal Aviation Regulations. Weather conditions below VFR minimums require flight under Instrument Flight Rules (IFR).

Essentially, VFR flight is based on the pilot controlling his airplane using visual references and cues outside the cockpit such as the nat-

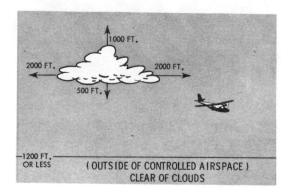

Fig. 7 Minimum distance from clouds—VFR, below 10,000 ft.

ural horizon or the ground. Flight instruments are only secondary references. When all references and cues outside the cockpit disappear, such as flying in a cloud, the flight instruments become primary in controlling the aircraft. Also, navigation and approaches to land must be accomplished by electronic means. IFR flight also requires control by the FAA Air Traffic Control (ATC) facilities to prevent collisions between airplanes.

In order to fly in IFR conditions, a pilot must have an instrument rating which requires additional written and flight tests.

LEARNING WEATHER

Learning weather is much the same as learning the other facts about flying. It takes time and application on the part of the pilot during his whole flying career. Nothing about flying is simple; all attempts to simplify it have shown that it cannot be done. However, all matters concerned with flying are within

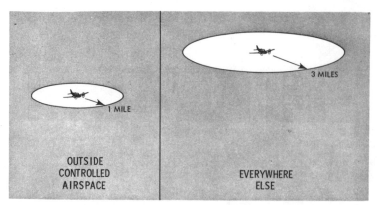

Fig. 8 Minimum visibility—VFR, below 10,000 ft.

the capabilities of a normal individual.

Skill in applying weather information cannot be taught in all its parts out of a book or in a classroom. A great deal about weather can be learned that way but not everything.

Experience in the air is still necessary.

A pilot, recently qualified, would be foolish to think that he had more than a foundation of weather knowledge. Experience, combined with endless study, must follow.

Fig. 9 Don't let weather surprise you.

Fig. 10 A typical weather observation console for measuring basic surface weather elements such as air temperature and wind velocity.

The sea of air in which we live and in which we fly is composed of many basic elements. These elements affect both you as a pilot and the airplane itself. A possible classification of these familiar elements of weather is:

- wind
- turbulence
- visibility
- icing
- hail
- precipitation
  - rain
  - snow
- fog

Only a few of these elements really affect the aerodynamics of the airplane. It flys equally well in thick clouds or clear air. Extreme turbulence and hail can endanger the aircraft and ice accretion reduces the aerodynamic efficiency of the airplane.

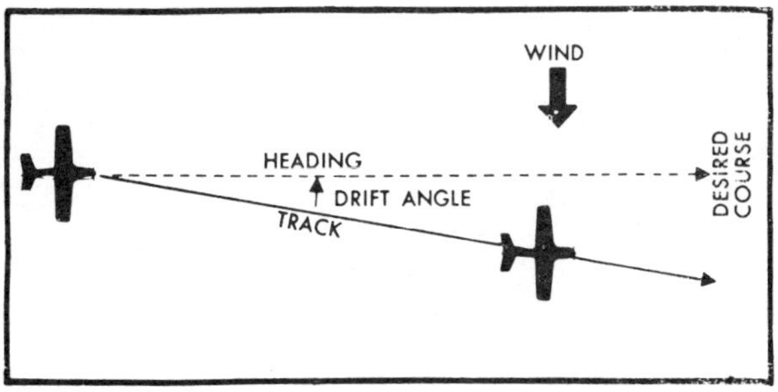

WIND

Wind influences your navigation, takeoff and landing and ground handling. For naviga-

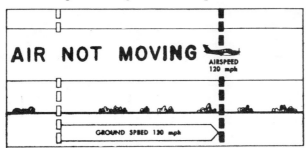

Fig. 11 A crosswind will cause the airplane to drift off-course unless a heading correction is made.

tion purposes, you must know the wind direction and speed at your cruising altitude. High winds on takeoff reduce the ground roll of your airplane and increase your angle of climb. On the other hand, a cross wind demands increased skill both on takeoff and landing. High surface winds also present problems in taxiing especially in light airplanes.

TURBULENCE

Turbulence affects both you as a pilot and the airplane itself. Turbulence, or rough air,

Fig. 12 With no wind, navigation is no problem.

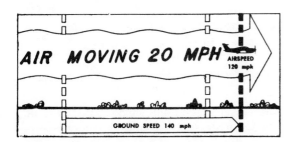

Fig. 13 A direct tailwind increases the airplane's groundspeed.

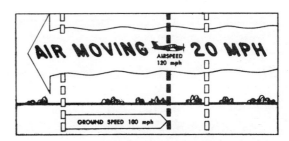

Fig. 14 A direct headwind decreases the airplane's groundspeed.

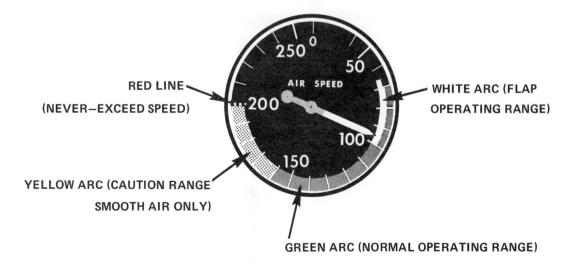

Fig. 15 *Federal Aviation Regulations require appropriate color-coding on the airspeed indicator. "Maneuvering speed", however, is not marked on the airspeed indicator. It is specified in the airplane flight manual.*

affects the comfort and well-being of both you the pilot, and to a greater extent, that of your passengers. The one element that can forever sour a non-pilot on flying is a long cross country flight in a light airplane in turbulent air. Besides the effect on comfort, turbulence makes control of the airplane more difficult, especially while flying on instruments. Holding airspeed, altitude, and heading during turbulent conditions can be a real chore.

Heavy turbulence can affect the structural integrity of your airplane. The degree of turbulence an airplane can withstand is a function of indicated airspeed. At the "maneuvering speed" as specified in the airplane flight manual, the airplane can withstand the most severe turbulence.

be capable of controlling his airplane entirely by instruments. Navigation must use electronic means. If you, as a pilot, cannot fly "instruments", visibility is an extremely important weather element. Even instrument pilots must have certain minimum visibility conditions during the approach and landing.

ICING

Most people believe that it is the *weight* of ice which adversely affects an airplane's performance. Although weight is a significant factor, especially in a light airplane, it is the reduction in aerodynamic efficiency that is of major concern. Ice on wings and control

Fig. 16 *Knowledge of weather can help you avoid turbulence.*

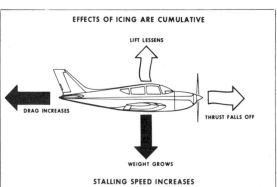

Fig. 17 *The effects of icing are cumulative, resulting in a total performance decrease.*

VISIBILITY

As previously stated, the airplane flies equally well in clouds or clear air. Visibility, however, has a profound effect on you as a pilot. Flight in clouds requires that the pilot

surfaces changes their shape. This adds up to loss of lift, increase in drag, and a change in aerodynamic balance of the airplane affecting its controllability. At the time wing and control surface icing occurs, propeller icing is

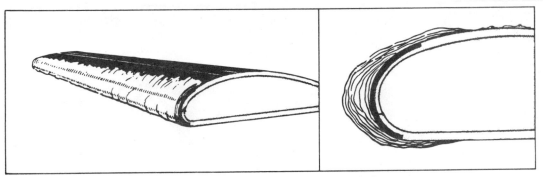

Fig. 18 Ice distorts the shape of an airfoil.

also likely, resulting in a power loss. It is readily apparent that the combination of all these factors can be disastrous.

HAIL

Although rarely encountered, hail can severely damage an airplane, even if it's sitting on the ground. Hail is normally associated with thunderstorms which should be avoided by *all* airplanes, not just light planes.

PRECIPITATION (RAIN, SNOW)

Unless icing occurs, the major effect of rain or snow is their effect on visibility. Except in extreme cases, the airplane will fly normally in a rain or snowstorm. However, it doesn't take much rain or snow to reduce visibility

to values below VFR minimums.

FOG

Again, an airplane will fly normally in fog. Since fog lies on the earth's surface, it can severely restrict takeoffs and landings. Even instrument pilots require certain minimum visibility values in order to accomplish a landing after an instrument approach. A heavy fog can close an airport to all operations for days.

SUMMARY

These are the familiar weather elements affecting flight. How, when, and where these elements occur, and to what degree and intensity, is of major concern to the pilot.

Fig. 19 Knowledge of weather is essential to both the VFR and IFR pilot.

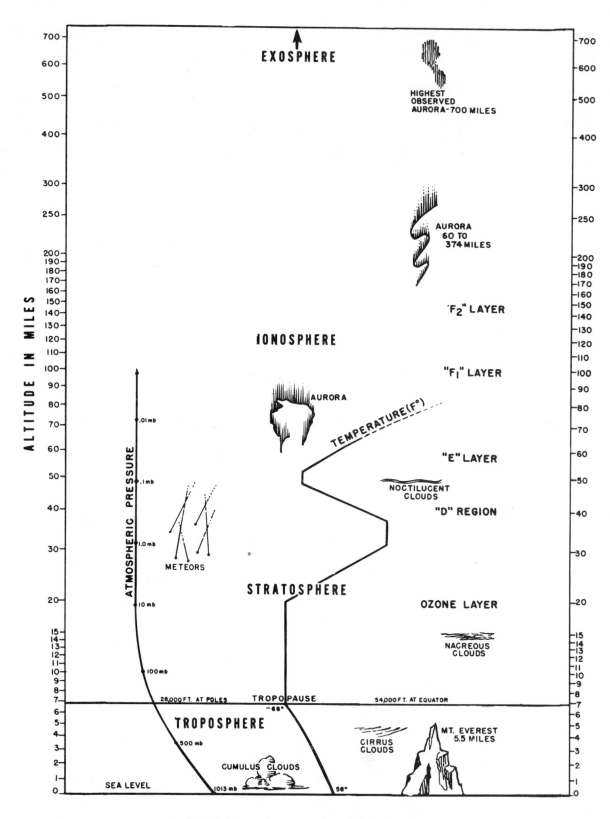

Fig. 20 Schematic cross-section of the atmosphere.

THE ATMOSPHERE 3

Because weather is the most important factor in the accomplishment of any flight, it is mandatory that pilots be thoroughly familiar with it. Safety of flight demands it.

It is not the purpose of this book to teach the principles of weather forecasting to pilots, new or old. Good weather analysis and forecasting require a highly technical knowledge of all phases of meteorology. Forecasting weather is the job of the men who have studied and learned of its vagaries. However, a pilot who possesses a basic knowledge of weather is better able to understand the weather briefing he receives. You will become a better-informed and more professional pilot if you realize this fact and get all you can out of all available weather data.

The mass of air surrounding the earth, covering land and sea alike, shares many of the characteristics of the oceans. Although we cannot see it, it is as real as the water that covers some three-quarters of the earth's surface. Like the oceans, the air rotates with the earth as the earth orbits about the sun. However, unlike the seas, the air has its own circulation relative to the earth's surface.

COMPOSITION AND PROPERTIES OF THE ATMOSPHERE

The colorless sea of air enveloping the earth is called the atmosphere. It is characterized by general motion (currents), and variations from the general motion. Some winds flow with the regularity of the great ocean currents, while others are as changeable as the flight of a bumblebee.

Ingredients of the Atmosphere

Air, the material of the atmosphere, is a chemical mixture of several gases. At sea level dry air contains about 21% oxygen, 78% nitrogen, and about 1% argon. Other gases such as carbon dioxide, helium, krypton, and neon are found in very small and varying amounts. Many scientists believe that the earth may be unique among the planets in having free oxygen in its atmosphere.

There is another component of air, although found in variable and relatively small quantities, that is of major importance in meteorology — water vapor. A molecule of water vapor weighs about g/8 as much as a molecule of dry air. Without water vapor in the air we would have few of the phenomena which comprise what we call "weather." There would be no cloud cover to protect us from the sun, no rain to water the plant life upon which animals and man depend for food. Life in its present form could not exist upon the earth.

The water vapor content of the atmosphere varies with latitude and altitude; it is lowest at high latitude and high altitude. Although practically all water vapor is concentrated below the 25,000-foot level, no air with a water vapor content of zero has been found in the atmosphere up to at least 250,000 feet.

Solid particles such as dust, smoke, and salt from sea spray are found in the lower layers of the atmosphere. They range greatly in size, from submicroscopic to those large enough to be seen with the naked eye. These particles reduce visibility and some act as nuclei for condensation of water vapor.

Although air is light, the weight of the atmosphere is enormous. However, man is not aware of this because the pressure of the air and fluids within his body counter-balances the pressure exerted on the outside of his body. The pressure exerted on the entire earth is 2116 pounds per square foot, or 14.7 pounds per square inch.

The evaluation of some elements or activities of the atmosphere such as turbulence, clouds, fog, haze, rain, and snow can be made by seeing or feeling. Others such as pressure, temperature, wind direction and speed, and relative humidity can be evaluated better by instruments.

Layers of the Atmosphere

In meteorology the earth's atmosphere is generally divided into two major regions — the lower atmosphere and the upper atmosphere.

The lower atmosphere is called the troposphere. The thickness of the troposphere fluctuates with time and latitude, but averages about 54,000 feet over the equator and about 28,000 feet over the poles. In the temperate zones seasonal changes greatly affect the

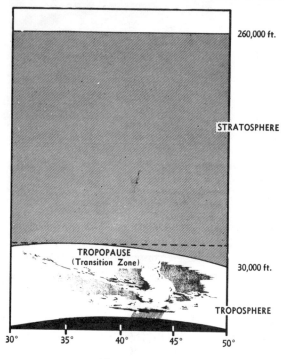

Fig. 21 Tropopause.

thickness of the troposphere. It is thicker in summer than in winter. The thickness of the troposphere is constantly changing because of the temperature changes of the earth and the atmosphere — from day to night, and season to season. Even with these seasonal and daily changes the thickness of the troposphere in the temperate zone is rarely as deep as it is in the equatorial areas or as shallow as it is in the polar areas. At 45° latitude the average height of the troposphere is approximately 35,000 feet.

The troposphere is probably the most unsettled layer of the earth's atmosphere. It is the region of most cloud activity and most visible weather. Here are spawned nearly all the air masses, fronts, and storms that give the earth its weather.

Most of the weather is confined to the troposphere for several reasons, namely:

    A high percentage of the water vapor is in the lower atmosphere.

    Condensation nuclei are concentrated in the troposphere.

    Heating and cooling by radiation are greatest there.

The troposphere is the area in which the temperature drops at an average rate of approximately 2°C/1000 feet of altitude. This average temperature decrease is called the Standard Lapse Rate. The altitude at which this temperature decrease changes significant-

ly is the zone or boundary separating the lower atmosphere from the upper atmosphere. This imaginary zone is known as the tropopause and it is often considered as a sharp, narrow zone of transition.

The average altitude at which the tropopause is found varies systematically with latitude, being higher at the equator and lower at the poles. The tropopause temperature at 45° latitude is an average — 55°C; at the poles the tropopause temperature is warmer than at the equator (due primarily to the fact that the troposphere is so much more shallow in the arctic regions).

Remember that the tropopause is just the dividing zone between the lower and upper atmosphere; its height will shift constantly with changes in the thickness of the troposphere.

The region of the atmosphere above the troposphere is commonly divided into layers or shells, according to several sets of criteria of which temperature distribution is the most common. The next layer above the troposphere, according to this criterion, is the stratosphere. The term stratosphere has been used to denote both (a) the relatively isothermal region immediately above the tropopause, and (b) the shell or layer extending from the tropopause to the minimum temperature level at about 250,000 feet altitude. The term mesosphere is now used to denote the shell which has a broad maximum temperature at about 160,000 feet altitude extending from the top of the stratosphere to about 250,000 feet. The term thermosphere is used to denote the shell above the mesosphere with a more or less steadily increasing temperature with height. Another criterion is the distribution of various physico-chemical processes. The ozonosphere is the term used to denote the first layer according to this criterion, lying between approximately 32,000 feet and 165,000 feet, and is the general region of the upper atmosphere in which there is an appreciable concentration of ozone. The ionosphere is the term used to denote the next layer under the physico-chemical process concept which starts at an altitude of about 250,000 feet.

Several other terms are used to distinguish still higher layers such as the chemosphere, the neutrosphere, the exosphere, the homosphere, and the heterosphere, but we are not concerned with them here.

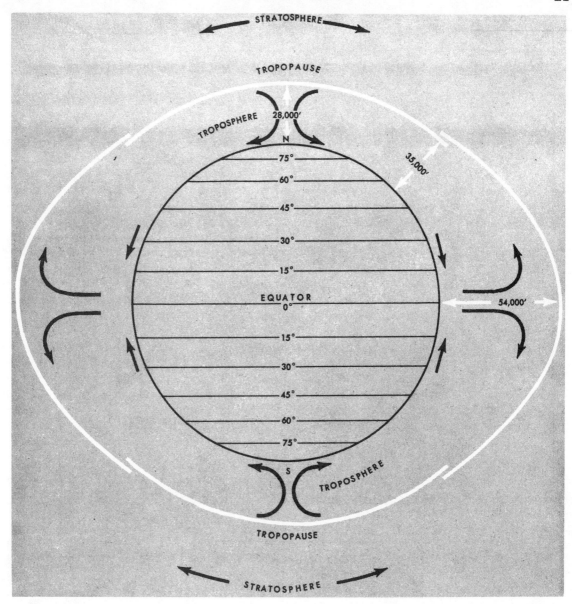

Fig. 22 *General structure of the atmosphere.*

4 HEAT ENERGY IN THE ATMOSPHERE

For million of years the sun has been radiating its light and energy to the earth from what seems to be a never-ending supply. The sun plays a tremendous part in our everyday living but we accept its warm rays and its life-sustaining energy without thinking.

RADIATION

Radiation is the process by which the sun's energy is propagated through space by virtue of changes in the electric and magnetic fields in space. In many ways, however, radiant energy acts as if it were transmitted in the form of waves. The wave point of view is often used to explain problems in weather.

Radiation occurs over a considerable range of wave lengths, depending on the temperature of the radiating body. The light we see and the heat we feel from the sun's surface is emitted as short wave radiation. Cooler objects, such as the earth, emit energy as long wave radiation. Consequently, two different types of radiational transfer must be considered. Incoming solar radiation, or insolation, from the sun, and terrestrial radiation from the earth.

The energy radiated by the sun is the major cause of weather phenomena on the earth. The sun might be said to inaugurate the circulation of the atmosphere which, in turn, causes changes in atmospheric pressure and thereby produces wind. It is important to know the effect of the sun's energy on our planet.

First, this energy must pass through the atmosphere which surrounds the earth. Different substances will absorb or reflect energy in different ways and varying amounts. Of the total of short wave radiation emitted from the sun that reaches the earth, the earth absorbs only about 47%. The accompanying diagram shows that the remaining radiation is reflected to outer space or is absorbed by the atmosphere (34% reflected to outer space plus 19% absorbed by the atmosphere). The clear air

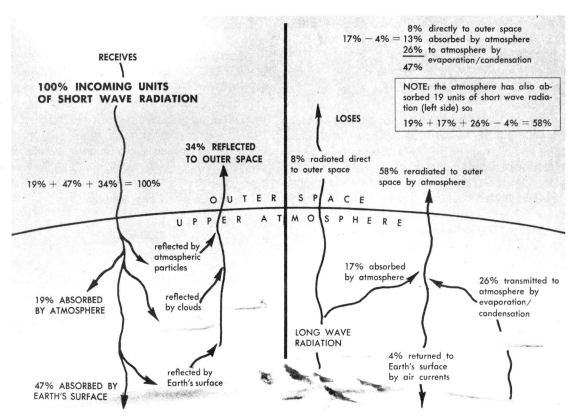

Fig. 23 Annual budget of heat loss and gain by radiation for the earth and atmosphere.

within the troposphere is nearly transparent to short wave radiation. In general, the short wave radiation from the sun passes on through the atmosphere and does not affect the temperature of the air to any great extent. In the early part of the morning when the earth starts to receive radiation from the sun, the temperature of the earth increases before there is an appreciable increase in the temperature of the air.

Four factors responsible for seasonal and geographical variations in general weather conditions on the earth are:
1. The earth's daily rotation about its axis.
2. Its yearly motion about the sun (revolution).
3. Nonuniform heating of the earth's surface.
4. Spheroidal shape of earth.

The heating of the earth during the day and cooling at night is primarily a result of the earth's rotation on its axis. As the earth turns from west to east, the side facing the sun is heated; when night comes, the same side is opposite the sun and will begin to cool. The ground and the air in immediate contact with it will generally reach its lowest temperature shortly before sunrise.

The effects of the earth's yearly revolution around the sun are modified by the tilt of its axis. In Figure 24, Variations in Solar Energy Received by the Earth, notice that the areas under the direct or perpendicular rays of the sun receive more heat than those under the slanting rays (comparing equal areas). The slanting rays also pass through more of the atmosphere which absorbs, reflects, and scatters the sun's energy. This accounts for the difference in the warmth of sunlight at 0800 local time when the rays are slanting, and at midday when they are more nearly perpendicular. Also, there is less radiation at a given area near the poles than near the equator because the earth is a spheroid.

Each year the perpendicular rays of the sun migrate from 23½° North Latitude (21 June) to 23½° South Latitude (22 December), causing the seasons of the Northern and Southern Hemisphere. The diagram, Effect of Inclination of the Earth on Seasons, Figure 25, indicates why the rays are perpendicular at 23½° North Latitude on June 21, at the equator on September 22, at 23½° South Latitude on December 22, and at the equator again on March 21.

Fig. 24 Radiational heating from the sun.

Unequal duration of daylight contributes to the uneven distribution of heat. The diagram also shows that each pole has 6 months of daylight and 6 months of darkness. On June 21, all territory within the Arctic Circle has 24 hours of daylight; on December 22, all territory within the Arctic Circle has darkness or twilight.

The nonuniform heating of the earth's surface is another factor that produces weather within the troposphere. This is caused by the different reactions to heat of land and water.

Absorption of heat energy from the sun is confined to a shallow layer of the land surface. As a result, the land will heat faster during the day and cool faster during the night than will the water surface. Water surfaces heat more slowly than land surfaces because:

The sun's rays can penetrate water better.

The movement of water distributes the heat over larger areas.

The specific heat of water is about three times that of land; that is, about three times as much heat is required to raise the temperature of a given mass of rise in temperature of an equal mass of water as is required to effect the same land.

Evaporation, which is a cooling process, occurs over water.

Vertical mixing occurs in water.

Color, texture, and vegetation influence the rate of heating and cooling of the ground. Generally, dry surfaces heat and cool faster than moist surfaces. Plowed fields, sandy beaches, paved roads, and runways become

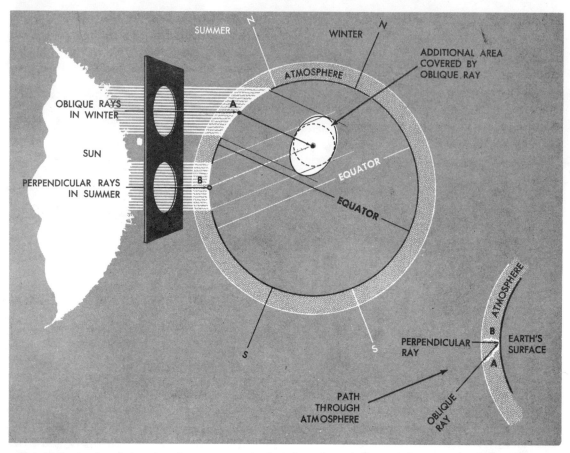

Fig. 25 Seasonal variations in solar energy received at the surface of the earth due to angle of the sun's rays.

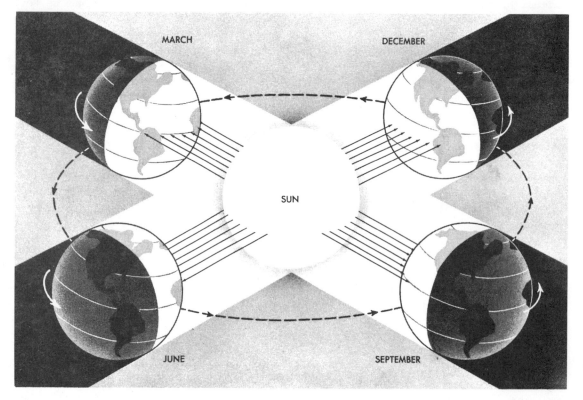

Fig. 26 The earth's seasons are caused by the inclination of its axes.

hotter than surrounding meadows and wood-
ed areas. During the day the air over a
plowed field is warmer than over a forest or
swamp; during the night, the situation is reversed.

Heat energy absorbed by the earth must be
reradiated out into the atmosphere so that a
heat balance will be maintained over the
years. This is accomplished by terrestrial radiation. The earth, being a relatively cool object, radiates its energy in the form of long
waves which are invisible to the human eye.
This goes on night and day, but the earth can
transmit energy at only a fraction of the rate
that it is received from the sun during the
day, so it is only at night that the effect of
the terrestrial radiation shows up as a net
cooling of the ground.

The atmosphere acts as a shield over the
earth by transmitting most of the solar radiation and absorbing or re-emitting the long-
wave terrestrial radiation. This restriction of
cooling is called the greenhouse effect, and
results in a heating effect on the atmosphere.
As we have already seen, the atmosphere is
nearly transparent to the incoming solar radiation, but in general the moisture in the air
is capable of absorbing long wave or terrestrial radiation. The ability of the atmosphere
to absorb terrestrial radiation is dependent
upon the moisture content. This accounts for
the large day to night temperature range over
desert areas, and the small range over humid
areas.

This absorption of terrestrial radiation by
the atmosphere traps, for a time, some of the
energy which would otherwise be more quickly lost to space, thus preventing the atmosphere from cooling as rapidly as it would if
it were drier. The water in the atmosphere has
the same effect upon maintaining a warmer
atmosphere as the glass in a greenhouse.
If the atmosphere were as transparent to terrestrial radiation as it is to insolation, the
average temperature of the earth would be
much colder than it is. Figure 23 shows how
the earth and its atmosphere are able to
maintain a heat balance because of the earth's
ability to re-radiate the energy it receives
from the sun.

To summarize then, the earth is heated during the day by solar radiation or insolation,
and is cooled by terrestrial radiation (day and
night).

Fig. 27 Transfer of heat by conduction.

CONDUCTION

Conduction is the process by which heat is
transferred through matter, without the transfer of matter itself. Some solid substances are
good conductors of heat while others are not.
When a silver spoon is heated at one end,
the other end soon becomes hot by conduction. When one end of a piece of wood is
heated, the other end remains cool. Silver is
a good conductor, wood a poor one. Like the
piece of wood, still air is a very poor conductor of heat.

Heat transfer by conduction is defined as
going from the warmer to the colder object.
On a sunny day the earth's surface is heated
by absorbing insolation. After the earth's temperature becomes higher than that of the
surface air, the air in contact with the earth
is warmed by conduction. At night, the process is reversed. The earth is cooled rapidly
by terrestrial radiation and then the air in
immediate contact with the ground is cooled
as it gives some of its heat by conduction to
the cooler earth. This process continues
throughout the night so that the ground surface and air are cooled and both remain
about the same temperature. Remember that
air is a poor conductor of heat; this change
in the temperature of the air would be effective only for a few inches above the surface
of the earth were it not for wind and turbulence which distribute the cooling to greater
heights (a few feet or a few thousand feet, depending on the wind strength).

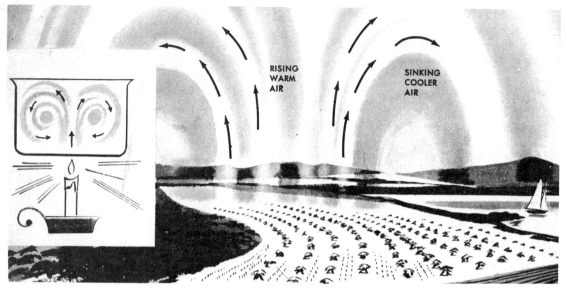

Fig. 28 Vertical currents (convection) produced by unequal surface heating.

The poor conductivity of the air and its slow loss of heat explains why ground frost can occur when the temperature at the standard 3 feet observation height is considerably above freezing. At night the grass, automobiles, aircraft, and other surfaces where frost forms, are often colder than the air a few feet above.

CONVECTION

As previously pointed out, various types of surfaces will absorb heat energy at different rates. For instance, air lying over a land surface will warm up (or cool off) faster than air lying over a water surface; air lying over a paved runway will heat faster than the air over the surrounding grassy areas. This unequal local distribution of heat will bring about another method of heat transfer, convection. In meteorology the term "convection" is used to indicate the transfer of atmospheric properties primarily by vertical motion. As the air is heated near the earth's surface, it becomes lighter or less dense than the surrounding air. The lighter air will rise, thus producing convection which is usually accompanied by turbulence. As a parcel of air rises the atmospheric pressure on it will decrease, producing expansion and cooling. Convection is often very noticeable along a coastline, especially during the summer, where the rising moist air will produce a line of cumulus clouds. The accompanying illustration, Figure 28, Vertical Currents (Convection) Produced By Unequal Surface Heating, shows how this process takes place.

ADVECTION

When air rises in convection currents, another method of transfer of an atmospheric property takes place, namely, advection. Advection is the transfer of some atmospheric property by horizontal motion of air (wind). When warm air rises in a vertical motion, the cooler surrounding air will move in and replace the air that has been lifted.

Temperature

According to the molecular theory of matter, all substances are composed of minute molecules which are in rapid motion among themselves. As the velocity of its molecular motion increases, the temperature of a body increases. The energy due to its molecular motion is called heat and is a measureable quantity. Air temperature is usually measured with a mercury thermometer.

Two scales commonly used for measuring temperature are the Centigrade and Fahrenheit scales. Pilots sometimes find it necessary to convert temperature readings from one scale to another for two reasons:

1. Surface temperatures are given in the Fahrenheit scale while upper air temperatures are given in the Centigrade scale.

2. Aircraft are equipped with Centigrade thermometers.

The accompanying diagram, Figure 31, includes the conversion formulas and a quick conversion scale. From the scale you will notice that 59°F is equal to 15°C.

As you gain altitude in an aircraft you will notice an over-all decrease in temperature.

Fig. 29 Convection currents from on-shore winds in daytime.

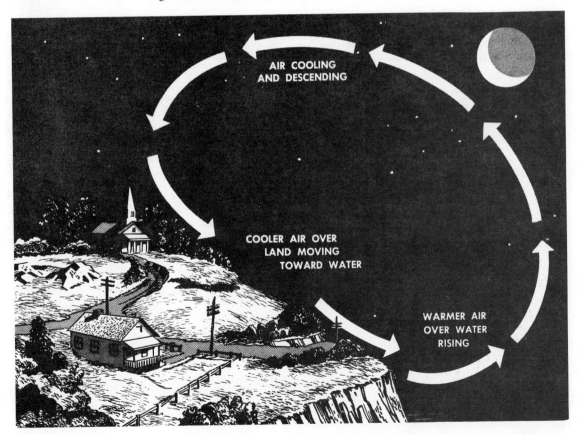

Fig. 30 Convection currents from off-shore winds at night.

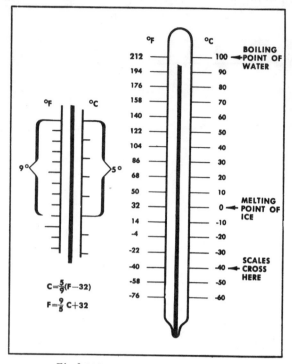

Fig. 3 1 Temperature conversion diagram.

This is due to the fact that the air nearest the earth is heated the most — it is closer to the insolation absorbing earth. The variation in temperature with altitude is expressed by meteorologists in degrees per thousand feet. This rate of cooling (lapse rate) varies from day to day and from one area to another. It depends upon the amount of heat energy reaching and escaping the earth, and upon vertical and horizontal atmospheric motions aloft. In the troposphere the average temperature decrease with altitude is about 2°C/1000 feet. Remember, this is an average and can vary greatly from one day to the next, as well as with place and height.

Inversions

Almost every day a layer of the atmosphere will be found that shows an increase of temperature with altitude, rather than a decrease.

this situation occurs frequently, but is generally confined to a relatively shallow layer. It is called an inversion.

Inversion occurs when a warm wind brings in air aloft warmer than the air near the surface; this is often called a frontal inversion.

The most frequent type of inversion over land is that produced immediately above the surface by nocturnal cooling of the earth's surface. The air near the ground is cooled by contact with the ground. If there is little wind to cause vertical mixing, the cooling does not extend very high. At an altitude of a few hundred feet the air temperature may often remain constant throughout the night.

Inversions are caused by descending air aloft which compresses and is heated while the air below remains stationary.

Inversions are formed by cold air moving under warmer air (a cold front); the associated inversion is often called a frontal inversion.

A warm wind blowing over colder ground or water forms an inversion a few hundred feet above the ground — called a turbulence inversion.

Another variation from the general rule of temperature decrease with altitude is the isothermal lapse rate. If the temperature does not change with a change in altitude, we have an isothermal (constant temperature) lapse rate (0°C/1,000 feet).

Restrictions to vision, such as fog, haze, smoke, and low clouds are often found in or below low inversions and isothermal layers. Another characteristic feature of inversions and isothermal layers is the absence of turbulence within them.

Inversions are especially important to a pilot since many land and sea fogs occur in their presence. Even without a thermometer, it is easy to determine the height of an inversion since rising smoke and dust will be stopped by the base of the inversion.

WATER IN THE ATMOSPHERE 5

Water, in its three states, solid, liquid and gas, is an important part of the atmosphere.

Water vapor, which is water in the gaseous state, is the most important single element in the atmosphere in the production of clouds and other visible weather phenomena. The water vapor in the air varies in amount and the amount is relatively small. Most of it is concentrated in the lower levels of the troposphere. If all the water vapor in the atmosphere were to fall in a single shower, it would cover the globe with a layer of water only one inch deep. Despite the small amount of water vapor in the air at any given time, it has a tremendous effect on weather phenomena.

The availability of water vapor for the production of rainfall largely determines the ability of a region to support life. The ability of water vapor to absorb and emit long-wave radiation is of primary importance in maintenance of the temperature balance. Without water vapor, the earth and lower atmosphere would have a much greater daily temperature range. On the other hand, water vapor

in the atmosphere may make flying hazardous when it changes into a liquid or solid state in the form of fog, clouds, freezing rain, hail, snow, or ice.

SOURCE OF WATER VAPOR

The oceans are the primary source of water vapor for the atmosphere. However, evaporation from lakes, rivers, swamps, moist soil, snow, ice fields, and vegetation also furnish the atmosphere with water vapor.

HOW TO EXPRESS MOISTURE CONTENT

The capacity of air to hold moisture is directly related to its temperature. In general, there is a maximum amount of water vapor that air can hold at a given temperature. The warmer the air the greater the maximum possible amount will be. Cooling the air will decrease the maximum possible amount. When air contains its maximum amount of water vapor it is saturated; when it contains less than its maximum it is unsaturated. Notice in the illustration, Figure 32, the difference in the amount of water required to produce saturation in warm air and cold air.

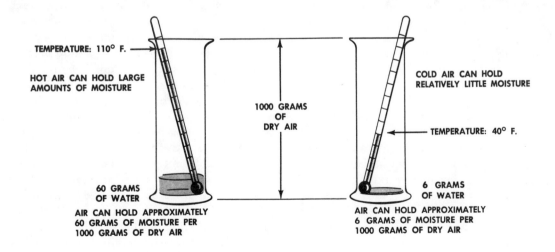

TEMPERATURE: 110° F.

HOT AIR CAN HOLD LARGE AMOUNTS OF MOISTURE

1000 GRAMS OF DRY AIR

COLD AIR CAN HOLD RELATIVELY LITTLE MOISTURE

TEMPERATURE: 40° F.

60 GRAMS OF WATER

AIR CAN HOLD APPROXIMATELY 60 GRAMS OF MOISTURE PER 1000 GRAMS OF DRY AIR

6 GRAMS OF WATER

AIR CAN HOLD APPROXIMATELY 6 GRAMS OF MOISTURE PER 1000 GRAMS OF DRY AIR

Fig. 32 Comparison of hot and cold air and effect on moisture content.

There are a number of terms used to express the water vapor content of air. Here we need discuss only two — dew point and relative humidity.

Dew Point

The dew point is that temperature, at any given pressure and constant water vapor content, at which the air will become saturated when cooled (when this temperature is below freezing, it is sometimes called the frost point). As the definition implies, the difference between the actual temperature and the dew point temperature, commonly called the spread, can be used to indicate, how close air is to being saturated. The closer the dew point temperature is to the air temperature, the greater is the likelihood of condensation and the formation of clouds and fog.

The dew•point temperature is also roughly indicative of the amount of water vapor present in the atmosphere; that is, the higher the dew point temperature, the greater the amount of water vapor in the air, for any given pressure.

Relative Humidity

Relative humidity is the ratio of the amount of water vapor actually present in the air to the maximum amount of water vapor the air can hold at a given temperature. This is expressed as a percentage and is another way of expressing the nearness of the atmosphere to saturation. For instance, when the air contains all the water vapor possible at a given temperature, it has a relative humidity of 100%. When it contains only one-half the maximum amount, it has a relative humidity of 50%. Occasionally, under some conditions, relative humidities greater than 100% exist in the atmosphere for short periods of time.

Condensation

There are two general processes whereby condensation is produced in the atmosphere: the addition of moisture to the air, and the cooling of the air.

When the water vapor in the air reaches 100% relative humidity, saturation occurs. Any further addition of moisture will cause some of the water vapor to change into the liquid form. This is called condensation. The most common forms of condensation are clouds and fog. If the temperature is cold enough (well below freezing) the water vapor will change directly into a solid form: ice crystals.

While it is true that the addition of water vapor to the air can produce saturation and condensation, saturation is produced more frequently by a process of cooling the air. After the cooling of the air has caused it to become saturated, any further cooling beyond saturation will generally produce condensation in the form of clouds or fog.

In the atmosphere this cooling may be produced by air passing over a colder surface (advection); by air being lifted, expanding and cooling (adiabatic cooling); or by the surface beneath the air cooling at night by radiation with resulting cooling of the lower layers of the atmosphere. This latter process most frequently produces fog rather than clouds, but winds of more than 15 knots may cause the fog to rise and become a deck of low clouds.

Adiabatic cooling (cooling by expansion) occurs as air is lifted to a higher altitude where the pressure is lower. The lift is provided by nature through thermal or mechanical means. Thermal lifting is the result of surface heating (convective lifting); mechanical lifting is the result of air being forced to rise over mountains (orographic lifting), over colder air masses (frontal lifting), or forced aloft by convergence (coming together) of air. Air moving over a colder surface can likewise be cooled to the extent that condensation will occur, usually in the form of fog or low clouds.

Fog and clouds are composed of small droplets of liquid water or of ice crystals. The small droplets of water form and collect on small particles of solid matter in the air such as dust, smoke, salt, or products of combustion. If it were not for these impurities (called condensation nuclei) in the atmosphere, a relative humidity of 400% might be required to produce condensation. These condensation nuclei are found throughout the lower layers of the atmosphere. Their abundance in the atmosphere permits condensation to occur generally at relative humidities around 100%. However, it is not uncommon to find some condensation (haze) in the atmosphere where relative humidities as low as 85% are being reported. This is explained by the fact that some condensation nuclei will begin to condense water at relative humidities well below 100%.

Liquid water droplets are frequently observed in the atmosphere at "freezing" temperatures, even as low as —40°C. This situation is called "supercooling," and is the prevailing situation in clouds down to a temperature of about —15°C.

Precipitation

Precipitation is visible moisture that falls from the atmosphere in the form of rain, sleet, snow, hail, drizzle, and combinations of these. Condensation does not necessarily produce precipitation, in fact most clouds do not precipitate. Initial cloud particles are usually very small and remain suspended in the atmosphere. Precipitation occurs when the cloud particles grow to a size large enough to fall freely in the atmosphere. This growth can occur through one or a combination of several processes. One process, which applies usually to clouds at or below freezing temperatures, requires the co-existence of supercooled water droplets and ice crystals. The ice crystals grow at the expense of the droplets and become snow flakes. If the flakes fall far below the freezing level they melt into rain drops. Collision of drops of varying size is another process that produces precipitation, and applies generally to clouds above freezing.

There are factors other than saturation and condensation which are required for the production of precipitation. The presence of vertical currents is one of these factors. Only very light rain, snow, or drizzle is found beneath clouds which do not have pronounced updrafts or general upward motion present within them, and usually rain of any considerable intensity requires the cloud to be over 4000 feet thick.

On occasions, when strong vertical currents are present in clouds, supercooled water droplets or ice particles in these clouds are carried to very great heights. These particles become very large before and during their fall, and thus reach the ground as heavy rain, snow showers, or even hailstones.

A note of interest is that all precipitation from clouds does not reach the earth, because on many occasions it evaporates completely in dry air below the cloud base before it can reach the ground.

6 STABILITY

The stability of the atmosphere determines how it will react to upward currents of air. In a stable atmosphere currents return to original position. An unstable atmosphere normally does not return vertically displaced air to its original position, and may actually speed up the displacement. To develop an understanding of atmospheric stability, we need to be acquainted with the mechanics of vertical motion, particularly motion upward. Several methods have been developed to check the atmosphere for stability; the most common, and the one we'll discuss here, is called the parcel method.

PARCEL METHOD

The parcel method requires that the density of a parcel of air at one level be compared with the density of air at another level. The parcel method permits a convenient comparison on thermodynamic or adiabatic diagrams, on which the temperature and moisture properties of the existing atmosphere are plotted. The changes an actual parcel of air would undergo while lifting occurs can then be seen on the chart.

Two other concepts need definition here, namely: lapse rate and adiabatic process. Lapse rate is the decrease of an atmospheric variable with height, usually temperature unless otherwise specified. Adiabatic process is defined as a thermodynamic change of state of a system in which there is no transfer of heat or mass across the boundaries of the

system. In an adiabatic process, compression results in heating; expansion results in cooling.

Atmospheric pressure decreases with altitude. In order to compare a parcel of air at one level with air at another level, the parcel has to be brought to the same pressure level as the air with which it is to be compared. The difference in density of the displaced air parcel and the air at the level to which it has been moved can be determined by the difference in the temperatures of the air parcels. Under constant pressure, the density of a gas, or mixture of gases such as the atmosphere, will vary inversely as the temperature varies; that is, the higher the temperature, the less the density (warm air is lighter than cold air).

After a parcel from one level (5,000 feet) rises to another level (10,000 feet), its temperature can be easily compared with the temperature of the air already at the other level (Figure 33). This comparison tells us how their densities differ. Remember that the effect of this rise is easily determined on special diagrams (thermodynamic or adiabatic diagrams). Therefore, we speak of lifting a parcel in the sense of testing what would happen to it if it were actually lifted in nature by passing up a mountain or front for example, or if it rose in a thunderstorm.

After either Parcel A or B is lifted to 10,000 feet, one of three situations occurs. In order to show these three conditions, let's

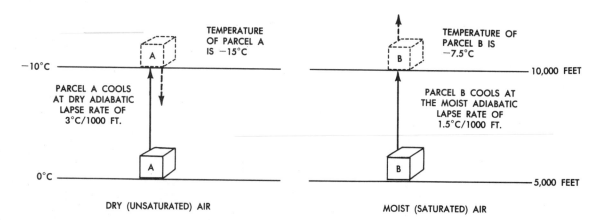

Fig. 33 Atmospheric stability.

visualize a roller coaster track and car.

Absolute Stability

Notice in Figure 34 that after a force is applied to the car and then removed, the car returns to its original position following a series of oscillations. The car resists displacement from its original position; it is in a state of absolute stability.

Neutral Stability

In Figure 35 a force is applied to the car, the car moves, but immediately stops when the force is removed. Note that the car remains in the new position; it has no tendency to return to its original posistion. It is said to be in a state of neutral stability, or neutral equilibrium.

Absolute Instability

The third condition of stability is illustrated in Figure 36. Note that once a force moves the car from its balanced position atop the

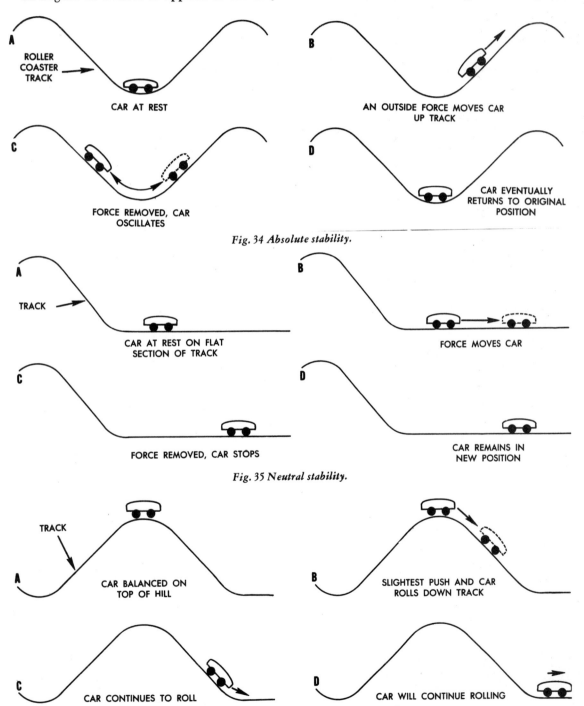

Fig. 34 *Absolute stability.*

Fig. 35 *Neutral stability.*

Fig. 36 *Absolute instability.*

track, the car will continue to move even though the force is removed. This situation is called absolute instability. (In this case keep in mind that it is atmospheric motion upward which produces clouds and weather phenomena, and in which we are most interested. The situations shown in Figures 34, 35, and 36 are only analogies.)

Now consider the situation of Parcel A and Parcel B in connection with the three conditions of stability just discussed. If the temperature of Parcel A, after being lifted to 10,000 feet, is less than the temperature of the air already at 10,000 feet, Parcel A will tend to sink toward the original level from which it started its upward motion. This would be absolute stability.

If Parcel A has the same temperature as the air at 10,000 feet after its lift from 5,000 feet, its density will be the same as the air at 10,000 feet and Parcel A will tend to remain at 10,000 feet. This situation is to be compared with the second condition — neutral stability.

If Parcel A has a higher temperature after its lift from 5,000 to 10,000 feet than the air at 10,000 feet, its density will be less than the air already at that level, and the parcel will experience a bouyant acceleration upward. This situation is analogous to the third condition of stability — absolute instability.

There is another aspect of the stability problem that must be mentioned, due to its importance in the formation of clouds. Note Parcel B in Figure 33. It is moist air, meaning that there is some water vapor (water in the form of a gas) contained in the parcel. In general, warm air can hold more water vapor than cold air, so that as soon as Parcel B begins its lift and cools, the amount of water vapor it can hold decreases. If during the lifting process the water vapor in the parcel becomes the maximum possible amount, the parcel is said to be saturated.

Any further lifting and cooling that occurs will cause water vapor to condense to liquid water. Condensation is a heating process which liberates the heat (approximately) that was originally required to vaporize the water. Therefore, the rate of cooling of a rising parcel in which water vapor is condensing will be less than the unsaturated-air rate. This, in turn, makes it easier for the parcel in which condensation is occurring to become warmer than the surrounding air, resulting in an acceleration upward. Thus, in general, instability is more easily achieved in moist air.

Thunderstorm Forecasting

It can be readily seen that the stability of the atmosphere plays an important role in the formation of vertically developed weather systems. Thunderstorms fall in this category, and thunderstorm forecasting techniques are based on the stability concept, though other factors, such as the forces available for lifting the air are equally important. Advanced concepts of stability have been developed to determine how much lifting would be required to change a stable air mass into an unstable one (potential instability). This information can then be correlated with other factors by the forecaster.

Virtually all weather problems that confront a pilot are associated in one way or another with clouds or fog and he should never fly into any cloud without knowing what weather elements he is likely to encounter and what he will do if they adversely affect his flight.

The production of icing conditions and precipitation is always associated with clouds. Turbulence, while not directly caused by clouds, may be caused by the same physical process which results in the formation of clouds, and there is the added complication that flight must be conducted on instruments in these conditions. A relatively slight accumulation of ice and moderate turbulence can have serious consequences when encountered in the course of instrument flight.

Cloud formations tell the pilot what the atmosphere is doing; they are visible evidence of motion and water content. Clouds are weather warnings to the pilot who can read them.

It is sometimes argued that a study of clouds, based on their appearance, is of no practical significance to the pilot. These ex-

ponents of "half-knowledge" say that the pilot needs to concern himself only with those characteristics of clouds which might adversely affect the normal conduct of flight. This is true as far as it goes. But knowing what will happen *after* one is in a cloud is a little like locking the barn door after the horse has been stolen. It might have been better to have avoided the cloud in the first place — if one had been able to identify it from its appearance.

Basically clouds fall into only two categories: cumuliform and stratiform. However, each category is divided into various cloud types determined by the height at which the clouds form and their content. Clouds are generally classified into the groups shown in Figure 37. The illustrations on the following pages show the different cloud types and the heights at which they occur. (Figures 38 & 39)

WHAT CAUSES CLOUDS?

Cloud masses of sufficient size and extent to warrant consideration as a flight problem may be formed by any one of three principal processes:

Family	Average Height	Stratiform	(Content)	Cumuliform	(Content)
High Clouds	20,000-40,000 Ft.	Cirrus Cirrostratus	Ice Crystal Ice Crystal	Cirrocumulus	Ice Crystal
Middle Clouds	6,500-20,000 Ft.	Altostratus Nimbostratus	Ice Crystal Ice Crystal	Altocumulus	Ice & Water
Low Clouds	near surface – 6,500 Ft.	Stratocomulus Stratus	Water Water	Stratocumulus	Water
Clouds with Vertical Development	1000-40,000 Ft.			Cumulus Cumulonimbus	Water Ice & Water

Fig. 37 General classification of clouds.

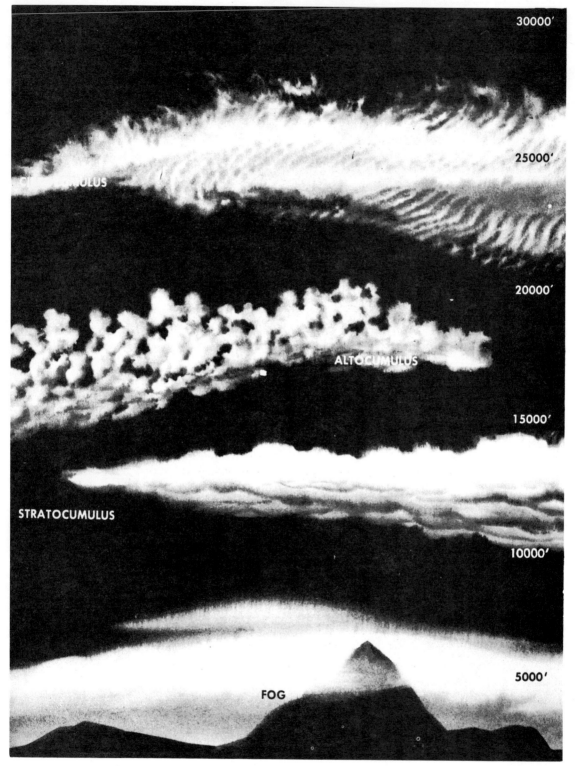

Fig. 38a Cloud forms.

1. Vertical convection.
2. Upslope motion.
3. Cooling by contact with a cold surface.

There are other processes of cloud formation but they are much less important from the pilot's standpoint.

CLOUDS FORMED BY VERTICAL CONVECTION

The usual cause of vertical convection is heating of the air at the ground level. Less usual causes are inflowing winds over a large area and lifting of conditionally unsta-

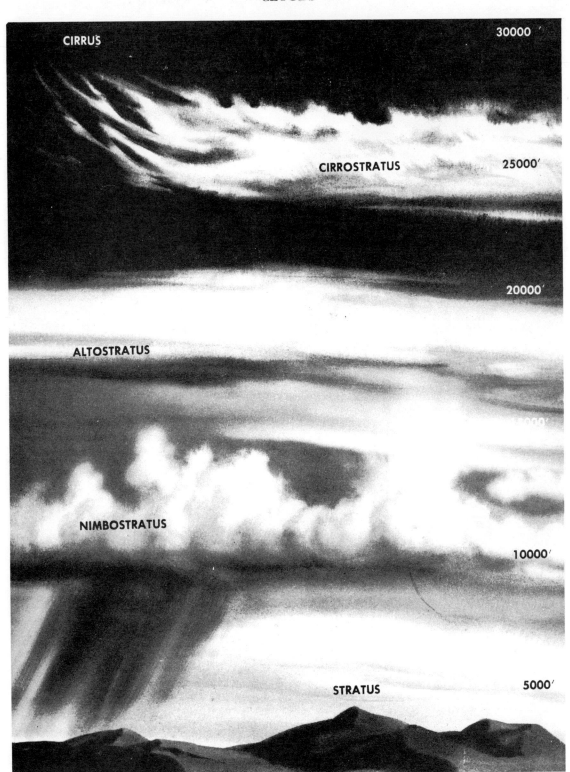

CIRRUS

30000′

CIRROSTRATUS

25000′

20000′

ALTOSTRATUS

NIMBOSTRATUS

10000′

STRATUS

5000′

Fig. 38b Cloud forms.

ble air. While heating at the ground will ordinarily result in clouds which build up from a low level, the last two causes may form convective clouds at intermediate or high levels. Most of the clouds formed by vertical convection are cumulus or cumulonimbus.

The individual rising currents of air which result from convection are surrounded by air which is sinking. The resulting clouds develop individually. This counteraction limits the amount of cloudiness that may develop as a result of vertical convection.

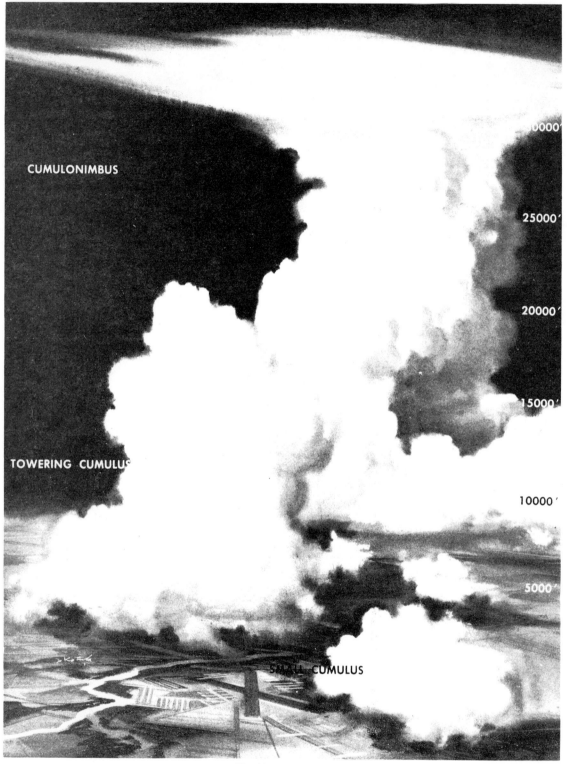

Fig. 39 Cloud forms.

While the sinking motion of the air around cumulus and cumulonimbus clouds is usually slow, an airplane will be obliged to maintain a climbing attitude in order to stay at a given level. The ground speed is thus reduced. If flight is made above the tops of the cumulus clouds, loss of ground speed from this cause will be avoided.

Because of the rapid cooling and condensation of water within a strongly developed cumulus or cumulonimbus, the amount of liquid water suspended in it is greater than in

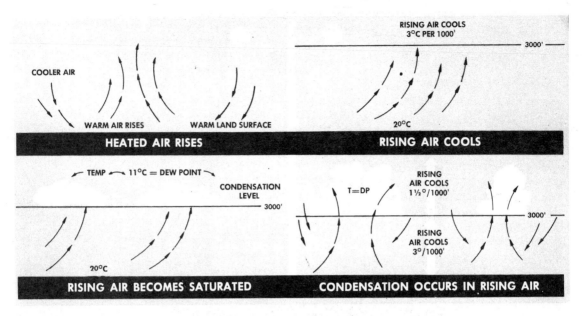

Fig. 40 Formation of cumulus clouds by convection.

any other type of cloud.
STAY ABOVE OR GO AROUND

Within deep tropical air masses, where high humidity is the rule and the air is conditionally unstable, heavy showers may occasionally fall from cumulus clouds whose tops are not above 8,000 feet. Vertical currents within such clouds may be strong and prolonged. The pilot should avoid these clouds by flying over or around them.

In large cumulus, the temperature within the cloud is higher, level for level, than in the surrounding air. Therefore, the elevation of the freezing level in a cumulus cloud is displaced upward, the average displacement being 500 feet or more. The displacement is greatest in the most active clouds.

This higher temperature within the cloud is of some importance in computing the lowest level at which ice will be encountered in cumulus or cumulonimbus clouds. Particular care must be exercised in cumulus clouds that are developing into squalls or showers because falling snow may contribute to ice formation in the portion of the cloud just above the freezing level.

Small cumulus clouds are typical of fair weather. Except for some turbulence and possible short periods of icing, they present no hazard to the pilot. They may be flown around with ease, topped at a reasonable altitude, or flown under, as the pilot chooses. His own comfort and the requirements of his flight are the deciding factors.

Cumulus clouds are primarily a daytime type, since they are usually the indirect result of solar heating. The occurrence of such clouds at night over land usually indicates conditions which may lead to widespread local storms. Whenever possible, it is best to stay out of cumulus clouds at night.

Over the oceans, the air during the day is normally warmer than the water; but at night, the water remains warm while the air becomes somewhat cooler. Surface heating of the air is thus most marked at night, and cumulus clouds over water accordingly develop mostly during the hours of darkness. When other conditions are favorable, clouds will form rapidly, and because of the high relative humidity of maritime air, the cloud cover is frequently unbroken and the base of the cloud low. Turbulence within such clouds is generally weak, but there may be local regions of more severe turbulence.

In northern latitudes, where the freezing level is within 2,000 feet of the surface, except during the summer months, the likelihood of encountering icing and cumulus clouds must always be remembered. The amount of ice formed in any one cloud may be negligible, but during a prolonged flight the accumulation can reach serious proportions. However, the depth of such cloud formations is usually not great and flight either on top or below the cloud base is both possible and recommended.

Fig. 41 Cumulus: A vertically developed cloud type which is usually dense and well-defined. The upper portions often resemble a cauliflower head, as shown above. Cumulus is composed of a great density of small water droplets, frequently supercooled. Larger drops often develop within the cloud and frequently fall from the base of the cumulus as rain or virga (trail of evaporating rain). Ice crystals may form in the upper regions of a large cumulus and tend to grow larger by taking water from the droplets. Cumulus occurrence often shows a diurnal variation being at a maximum frequency and development over land surfaces in the afternoon, and over water during the night.

CLOUDS FORMED BY UPSLOPE MOTION

If air moves up a long slope, its temperature is reduced by expansional cooling and the water vapors within it must eventually condense. The form of the resulting cloud, its vertical thickness, and its horizontal extent will be determined by the steepness and extent of the slope, the speed of the upslope flow, and the stability of the air. Sloping surfaces most important in causing upward flow of air are:

1. Warm fronts.
2. Cold fronts.
3. Sloping terrain.

The most important factor in determining how fast a cloud mass will develop within the upward moving air is the stability of the air. If it is conditionally unstable, deep clouds will develop rapidly soon after the first condensation occurs. The degree of turbulence and the intensity of icing will likewise depend upon the amount of instability. In stable air, the cloud will thicken more gradually.

Consideration must also be given to the fact that wind velocity may be greater aloft than at the ground. When this is the case, the higher air is lifted more rapidly than that be-

low and clouds, usually stratiform, form first at a high level.

UPSLOPE CLOUDS YOU WILL MEET

The principal cloud types formed as a result of upslope motion are:

1. Altostratus.
2. Altocumulus.
3. Stratus and nimbostratus.

Clouds formed by gradual lifting seldom produce noticeable turbulence unless the air is conditionally unstable. At times, particularly along fronts, friction with the underlying surface will produce choppy or bumpy air. Sustained vertical currents are exceptional, and if encountered, are invariably associated with cumuliform clouds.

With up-slope motion, the interior of a thick cloud layer may occasionally be quite turbulent, even though the cloud has no cumulus appearance. Drafts, choppy air, and side buffeting may be experienced at irregular intervals, and if precipitation is encountered, mild downdrafts may occur. Instrument flight under these conditions requires the full attention of the pilot. He should utilize the recommended procedures for flight in turbulence. Probably the element of surprise is the greatest hazard when such turbulence

Fig. 42 Cumulonimbus: This cloud is the ultimate manifestation of the growth of a cumulus and often extends from the lower levels of the atmosphere (2,000 to 4,000 feet) through the middle cloud altitudes up into the high cloud region in the vicinity of the tropopause (occasionally cumulonimbus grow well into the stratosphere). Cumulonimbus is (by definition) accompanied by lightning, thunder, and sometimes hail. Tornadoes (funnel clouds) are columns of violently rotating air pendant from cumulonimbus clouds.

is encountered. The forecaster can help in choosing the best flight levels. If the front slopes upward in the direction of flight, smoother air can usually be found at a lower level; if the slope is downward in the direction of flight, smooth air should be sought by climbing.

YOUNG CLOUDS BEAR WATCHING

Factors in estimating the hazard in up-slope forms of clouds include the age of the cloud mass, the rapidity with which it formed, and the time needed to fly through it.

When a cloud forms rapidly, reaching large size while it is still young, the rapid ascending motion of the air carries along a large amount of free water. When the temperature is below freezing, glaze will usually be encountered. The rate of ice accretion depends

Fig. 43 Cumulonimbus tops: A species of cirrus clouds which spread out from the cumulonimbus top into an anvil-shaped extension. This is a dense cirrus cloud composed of both supercooled water droplets and ice crystals, but mainly the latter. See also cumulonimbus under low clouds because the upper portions of these clouds extend well into the high cloud regions.

Fig. 44 Altostratus: A gray or bluish sheet or layer of cloud smooth in appearance, but also often associated with altocumulus. Altostratus often covers the entire sky and is frequently composed of a mixture of supercooled water droplets and ice crystals. Precipitation often falls from altostratus although it is usually light and of a relatively continuous nature. The sun or moon may be dimly visible through altostratus.

on the density of the cloud and the speed of the aircraft.

Old clouds, particularly those moving from a warm to a colder surface, generally lack turbulence. Much of the water initially contained in the clouds is lost as rain, so the hazard of ice accretion is much reduced.

The amount of ice accumulated in flying through clouds naturally depends on how far flight is maintained through them. To reduce the ice hazards, the flight path selected should be the shortest path through the clouds.

At high altitudes, with the temperature below —20°C, almost no ice is found except in well-developed cumulonimbus clouds. In flying through upslope clouds, the pilot should seek high altitude and low temperatures if the freezing level is close to the ground. However, unless the altitude of the cloud top is known, climbing through an icing condition should not be attempted, particularly if rain or snow is falling.

CLOUDS FORMED BY CONTACT COOLING

Whenever warm moist air moves over a colder ground or water surface, cooling of the

Fig. 45 *Altocumulus: Consists of white or gray layers or patches of solid cloud often with a waved aspect, the elements of which appear as rounded masses, rolls, etc., (occasionally also called mackerel sky). Small liquid water droplets usually supercooled, invariably comprise the major part of the composition of altocumulus.*

Fig. 46 *Stratus: A cloud type in the form of a gray layer with a rather uniform, structureless base, stratus usually does not produce precipitation, and is often formed by the dissipation or lifting of the lower layers of a fog bank. Stratus is composed of minute water droplets, or if the temperature is low enough, partly of ice crystals. In many areas this cloud type shows a daily variation and often repeats this variation cycle for several consecutive days, being lower and more widespread over land during the night and early morning.*

Fig. 47 Nimbostratus: A gray or dark massive cloud formation often diffused by more or less continuous precipitation of rain, snow, sleet, etc., and not accompanied by lightning, thunder, or hail. The precipitation usually reaches the ground. Nimbostratus is composed of suspended water droplets, sometimes supercooled, and of falling raindrops and/or snow flakes. Nimbostratus is often thick vertically and extensive horizontally, and totally obscures the sun, as shown in Fig. 47.

Fig. 48 Cloud formation over Stockton Field, California. Alocumulus and altostratus mixed to the right, with lowering altocumulus and stratocumulus beneath to the left.

Fig. 49 Mammatus cloud structure.

air near the surface will eventually bring about condensation of the water vapor in the air, forming clouds. The type of cloud will depend upon the wind velocity, the stability of the air, and the presence or absence of precipitation. Stratus is the most common type of cloud formed by contact cooling. Since warm air over a cold surface is cooled from below, the lowest air becomes heavier, increasing the stability. When the air is cooled below the dew point, condensation occurs and clouds begin to form. Since cooling progresses upward from the surface, stratus clouds are ordinarily low and shallow but they may spread over a wide area.

If the wind is strong, turbulence produced by friction will carry the cooling effect upward to a somewhat higher level. The cloud base will then be relatively sharp and definite and the cloud mass as a whole will tend to become stratocumulus.

Precipitation in the form of rain or snow seldom falls from stratus clouds. Light rime icing may be encountered, but it seldom reaches serious proportions except in continued flight without use of de-icers. Since the cloud is usually shallow, flight can be made above it without difficulty. Flying on or below stratus clouds is not recommended because the cloud base is low and even small hills protrude into it.

EXPECT LIGHT DRIZZLE

Unless the wind is strong the stability of the air suppresses turbulence. In light winds the base of the cloud is soft and hazy. It may be quite irregular in height, frequently varying several hundred feet within a short distance.

Stratus clouds with a light wind seldom produce icing conditions. However, drizzle may occur; and if the temperature is below freezing, glaze will accumulate at a slow rate. Flight should not be prolonged through freezing drizzle or even in the clouds above. In the course of a prolonged flight the accumulation of ice may become heavy enough to prevent the plane from climbing on top. If an instrument approach is made through a stratus layer and freezing drizzle is encountered, icing of the windshield will usually obstruct forward vision. Therefore, whenever a descent is made through a stratus cloud, even though the temperature on top is above freezing, it is wise to have the windshield de-icer operating. If there is none, open the forward window. Do not depend on a side window,

Fig. 50 Cirrostratus with cirrus above.

Fig. 51 High clouds are cirrus cumulus with a solid layer of stratus below.

Fig. 52 Stratocumulus with low ragged clouds of bad weather. Moderate shower visible in center of picture.

or "stick your neck out." The altimeter, air speed, and other instruments cannot then be watched; and they will be needed, particularly if scud clouds are present, until your wheels are on the ground.

CLOUD FLYING

It is well to remember that a low ceiling and low visibility do not in themselves demand a special flight technique. In the final analysis, the pilot has only to determine whether or not a landing can be made at a terminal. If a landing cannot be made, he must proceed to another field where the ceiling and visibility are adequate. A pilot who knows his clouds is prepared for the possible use of an alternate airport. Repeated attempts to land under unfavorable conditions only consume valuable fuel and increase fatigue, not to mention the danger of taking unwarranted chances. It is better to swallow one's pride and proceed to an alternate field after one unsuccessful attempt, or before, if conditions are known to be adverse.

8 · PRESSURE AND WIND

Wind is of vital concern to pilots. Upper winds affect flying range, groundspeed and aircraft headings, while surface winds determine landing and take-off conditions. Since the winds transport heat and water vapor in the atmosphere, they have an important effect on the formation of fog and clouds, the production of precipitation, and the horizontal and vertical distribution of temperature and water vapor. Wind is also the main agency through which the mass of the atmosphere is redistributed, thus causing the pressure to change.

Wind is air in motion; it occurs because there are horizontal pressure differences in the atmosphere. Since pressure is influenced by temperature distribution, wind is related to temperature. In fact, pressure, temperature, and wind are so interrelated that it is difficult to study them separately. In this chapter we will briefly discuss pressure, the relationship between pressure and temperature, and the relationship between pressure and winds.

Fig. 53 Pressure equal in all directions.

ATMOSPHERIC PRESSURE

Pressure is defined as force per unit area. Atmospheric pressure is the force exerted by the weight of the atmosphere above a given area. Observe how the force of the air is being measured in Figure 54. Thus, the pressure at a specified altitude is the weight per unit area of the atmosphere above this altitude. The average pressure at the surface of

the earth is approximately 2,116 pounds per square foot or 14.7 pounds per square inch.

Atmospheric pressure is continually changing. It varies with both time and location. These pressure changes are caused primarily by changes in the air density (weight of air per unit volume) produced by variations in the distribution of temperature.

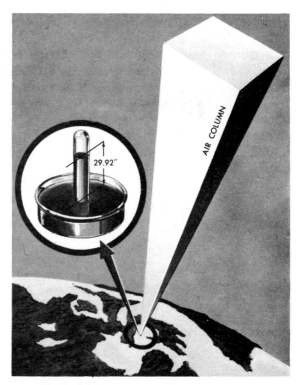

Fig. 54 Diagram comparing the weight on a column of the atmosphere with the weight of a column of Mercury.

When air is heated locally it rises until it reaches an altitude where its temperature is the same as that of the surrounding air. At that level it tends to spread out horizontally. This replacement of the cool air by the rising heated air reduces the weight (amount) of the air above the area being heated. On the other hand, the weight of air in adjacent areas is usually increased by the horizontal inflow aloft of air from over the heated areas. Thus, we expect to observe a decrease in pressure over regions receiving intense heat from the sun, and a rise in pressure over the adjacent regions.

Similarly, decreasing temperatures through a column of air will contract the column, producing lower pressure aloft and an inflow of air aloft into the column. When this happens, the surface pressure of the colder area is increased, and in the surrounding area pressure is reduced. We often observe an increase in surface pressure in cold land areas during the winter. The change in pressure in a particular location at any level is the sum of the variations of the pressures at all levels above it.

THE VERTICAL DISTRIBUTION

When you consider the fact that the weight of the sea of air which surrounds the earth is concentrated near the surface of the earth (due to gravity), it becomes obvious that atmospheric pressure decreases with increasing altitude. Pressure decreases very rapidly with altitude in the troposphere. In the stratosphere, the rate of decrease of pressure with altitude diminishes rapidly and becomes almost constant in the upper regions of the atmosphere.

Like most meteorological elements, the rate of decrease of pressure with altitude is not constant in the troposphere. It changes with both time and place. This variation is closely related to the fluctuation of the vertical temperature distribution (and more indirectly and complexly to the wind) in the troposphere. The rate of decrease of pressure. with altitude is greater in cold layers of the troposphere than in warmer layers.

Methods of Measuring Pressure

There are three methods of measuring atmospheric pressure. The most accurate and widely used is that of balancing the weight of the atmosphere with a column of liquid, usually mercury. The second method consists of measuring the response to variations in atmospheric pressure by an elastic diaphragm which encloses a partially evacuated chamber.

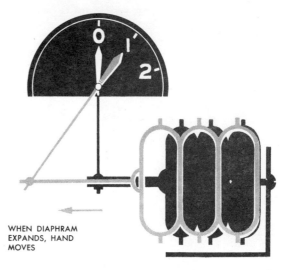

WHEN DIAPHRAM EXPANDS, HAND MOVES

Fig. 55 Diagram showing operation of an aneroid barometer.

The third method involves measuring the boiling point of a liquid. Since the boiling point of a liquid depends on atmospheric pressure, it can be used to measure the atmospheric pressure. The instrument used for making pressure measurements is called the barometer. One design that permits continuous recording is called a barograph.

The Mercurial Barometer

The mercurial barometer consists of a glass tube which is filled with mercury. This tube is open at one end and closed at the other. The open end of the glass tube filled with mercury is inverted into a vessel containing mercury. The column of mercury inside the glass tube adjusts itself so that its weight is equal to that exerted by the atmosphere on the free surface of the mercury. The height of the column of mercury is measured by a scale placed alongside the tube in the same fashion as a thermometer scale. The height of the column of mercury at any instant is directly proportional to the atmospheric pressure.

In some fields of science it is desirable to

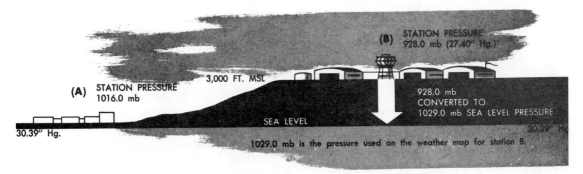

Fig. 56 Conversion of station pressure to mean sea level (MSL) pressure.

indicate atmospheric pressure in units of pressure (weight per unit area) rather than units of length (inches). In the metric system of measurement, a bar is a unit of pressure. The millibar (MB) (1/1000 of a bar) is used in meteorology to designate the value of atmospheric pressure. The conversion from one system of measurement to another is easily accomplished by the use of appropriate tables of conversion factors (e.g., 1 millibar equals approximately .03 inches of mercury). When the atmospheric pressure is 14.7 pounds per square inch, it supports a column of mercury 29.92 inches high. The corresponding pressure in millibars is 1013.25.

Although the mercurial barometer is the most accurate type and is used as a standard for measuring pressure, it has several disadvantages. Because of its size and sensitivity, it must be firmly mounted and regularly cleaned and calibrated. Most models of the mercurial barometer do not produce a continuous record of pressure. The aneroid barometer was designed to overcome this difficulty.

Fig. 57 Pressure decreases with altitude.

The Aneroid Barometer

Numerous types of aneroid barometers have been developed for various purposes, but all work on the same basic principle. They consist generally of a small, sealed metal box or chamber partially evacuated of air and flexible enough to respond to changes of air pressure. These changes in pressure are amplified and transmitted by a linkage system to a pointer or recording unit. Aneroid barometers may be calibrated to read in millibars, inches, or other desired units.

Because of their relative inaccuracy, aneroid barometers must be compared to mercurial barometers and corrected by adjusting a zero setscrew. Their main advantages are their relatively small size and their portability.

Pressure variations with both time and location are intimately related to weather elements (such as wind, clouds, and fog). Therefore, much attention is given to both surface and upper-air pressure so that they can be measured accurately and variations noted. In order to study and compare pressures from various locations, common reference levels must first be selected.

Mean sea level (MSL) is used as a reference level for surface pressure observation. If this MSL reference were not used, Denver would always have lower pressure than Miami regardless of the day to day variations in air mass characteristics. Surface pressure observations made at stations with different elevations are adjusted by use of air temperature and special tables to indicate what the surface pressure would be if these stations were located at mean sea level. In order to be of greatest value to the meteorologist, these observations must all be made at the same (Greenwich) time at all stations. Then, the sea-level pressures are plotted on weather maps for study and comparison.

That part of the atmospheric pressure which is produced by the pressure of the water vapor in the atmosphere is called vapor pressure. Vapor pressure is used to determine some performance characteristics of aircraft.

Pressure Patterns

A line which connects points of equal values of pressure is called an isobar. Isobars are drawn on surface (MSL) weather maps for selected intervals, usually every 4 millibars. These isobars outline pressure areas in somewhat the same manner as contour lines outline terrain features. The five following types of pressure areas, sometimes called pressure patterns or pressure systems, are outlined by isobars on surface maps.

Low: A low is a center of low pressure surrounded on all sides by higher pressure.

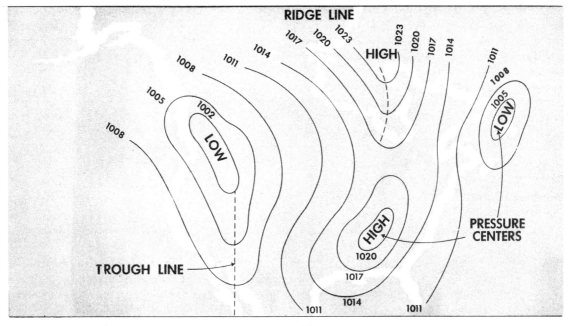

Fig. 58 Pressure patterns.

High: A high is a center of high pressure surrounded on all sides by lower pressure.

Col: A col is a saddleback region between two highs and two lows.

Trough: A trough is an elongated area of low pressure, with lowest pressure along the trough line.

Ridge: A ridge is an elongated area of high pressure with highest pressure along the ridge line.

Examples of these systems are found in Figure 58 and 59.

CONSTANT PRESSURE CHARTS

Forecasters need a three dimensional picture of the atmosphere and, therefore, require charts at various levels as well as at the surface. We know that pressure varies from time to time and place to place at sea level. It follows that pressure must also vary at any given altitude. Figure 62 shows an example of horizontal variations in pressure at a constant altitude of 10,000 feet. Notice the location of the high-pressure area at 10,000 feet.

The heavy line represents the position in a vertical cross section of the 700 MB constant-pressure surface. That is, this is a surface where the pressure is equal to 700 MB at all points, but whose true altitude varies. Notice that where low pressure existed at the constant altitude of 10,000 feet, low height above MSL now exists at the 700 MB constant-pressure surface.

Where high pressure existed at the constant altitude surface, high height exists at the 700 MB surface. Meteorologists have found it more convenient to use constant pressure and variable heights on upper-air charts than variable pressure and constant height. However, it is more useful and convenient to use a constant height (MSL) and variable pressure for surface charts.

Upper-air constant-pressure charts are frequently constructed by international agreement with reference to the following constant-pressure levels:

PRESSURE LEVEL (Millibars)	APPROX. ALTITUDE (Feet MSL)
1,000	400
850	5,000
700	10,000
500	18,000
300	30,000
250	34,000
200	39,000
150	45,000
100	53,000
50	68,000
25	82,000

The true altitude of a constant-pressure level varies from day to day and place to place. The heights as listed in the table for constant pressure levels are average values. The basic data plotted on the upper-air chart is the computed true altitude of the appropriate pressure. Lines which connect points of equal height (contours) are drawn on each chart for

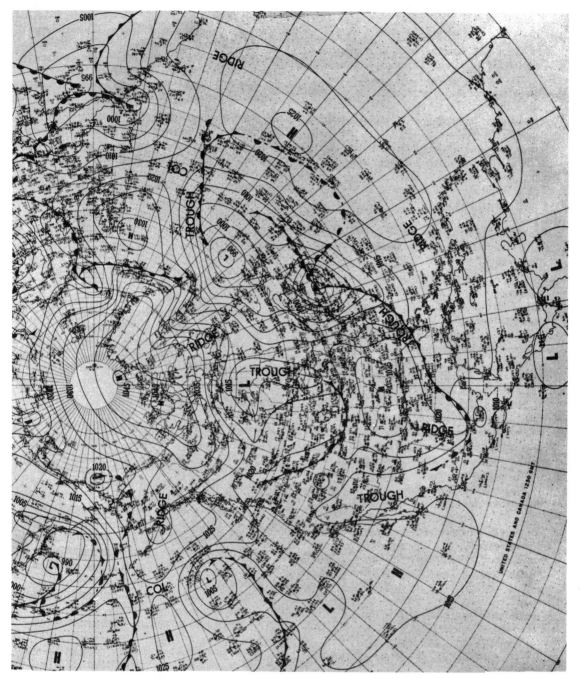

Fig. 59 Surface map with pressure pattern features marked.

selected intervals. The contours reveal the areas of relatively high and low heights of the given pressure chart. These areas of high and low height have a similar meaning on constant-pressure charts as areas of high and low pressure on surface weather charts.

The 500 MB constant pressure chart in Figure 61 is valid for the same time as the accompanying surface chart in Figure 59. By comparing these one can see how pressure systems change in pattern and intensity between the surface and the 500 MB level.

UPPER-AIR OBSERVATIONS

We need observations aloft in order to construct the upper-air constant-pressure charts. An instrument, called a radiosonde, is used to obtain data on the vertical distribution of pressure, temperature, and water vapor. The true altitude of the various pressure levels is computed from this information. The radiosonde consists of a small radio transmitter and a modulator containing elements sensitive to changes in pressure, temperature, and water vapor content. As it is borne aloft by a

Fig. 60 Weather relief map.

gas-filled balloon, an electronic ground station collects and records the data sent out by the transmitter. Figure 63 shows a radiosonde observation being made.

Most radiosonde stations are equipped to make winds-aloft observations along with the radiosonde observation. These observations are called rawinsondes. The wind information is obtained by tracking the radiosonde with a radio direction-finding unit. Wind speed and direction at various levels are computed from the observed azimuth and elevation angles. Sometimes wind data is obtained by radar from a reflecting target carried aloft by a balloon.

Additional winds-aloft data are obtained by stations equipped to make pilot balloon observations (PIBAL). The ascent of a gas-filled balloon which rises at a predetermined rate is observed through a theodolite. As in the tracking method of the radiosonde, wind direction and speed at various altitudes are computed from the observed azimuth and elevation angles. The obvious disadvantage of this method is that the balloon can be obscured by clouds and/or poor visibility. This, of course, results in frequent terminations of PIBAL observations at low altitudes. Upper-air observations are made simultaneously at selected land stations and specified ocean vessels in the Northern Hemisphere at least twice daily at fixed Greenwich times. Information is collected, evaluated, and exchanged on an international basis.

Over unpopulated areas, such as oceans and deserts, upper-air weather data is obtained by weather reconnaissance aircraft which make scheduled flights and release dropsondes (radiosondes dropped on parachutes) at selected intervals. Other types of upper-air observations which are used include wiresondes (kytoons), rocketsondes, and a high-altitude sounding device consisting of a balloon and rocket combination called a rockoon. Even though forecasters have all these means of obtaining weather data, the devices are expensive and do not cover all desired areas, altitudes and times. Therefore, great dependence is still placed on pilot-weather reports to supplement the upper-air observations.

Cold Core Highs

A high-pressure area that has colder temperatures at or near its center than the temperatures of the surrounding area (say 300 or more miles away) is called a cold core high, or simply a "cold high." The vertical separation or thickness of the layer between any given two constant-pressure surfaces is directly proportional to the mean temperature be-

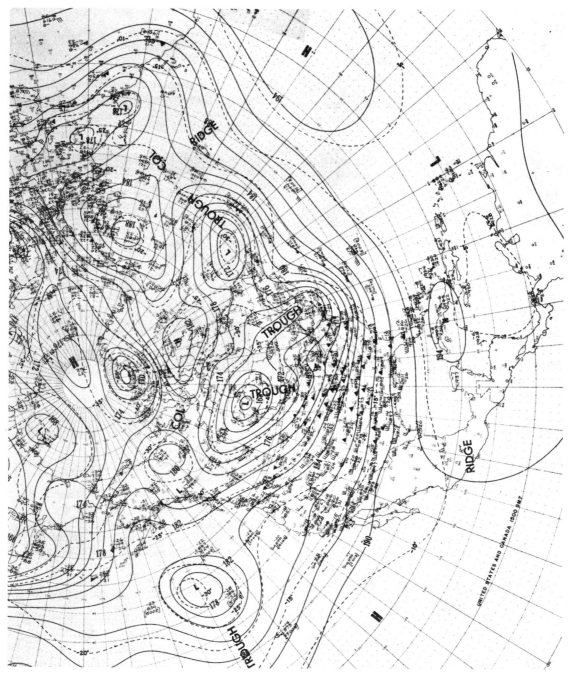

Fig. 61 Upper air constant pressure chart with height pattern features marked.

tween these surfaces. In the diagram (Figure 64) there is a cold core high at 850 MB. This high has disappeared at 700 MB and is a low at 500 MB.

In Figure 59 the large surface high in northern Canada is an example of a cold core high. Notice that directly above this high there is a very pronounced low on the 500 MB chart. A cold core high decreases in intensity aloft, since the cold temperatures which are favorable for a surface high are

also favorable for a low aloft.

Warm Core Highs

A high pressure area that has warmer temperatures at or near its center than around the periphery is called a warm core high. In Figure 58 the large subtropical high near 30° North, 40° West is still large and even more intense at 500 MB. Observe that warm core highs increase in intensity aloft and slope toward warmer air with increasing altitude. That is, the center of the high at 500 MB is

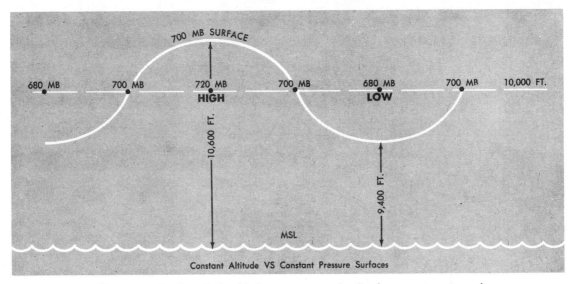

Fig. 62 *Diagram showing relationship between constant level and constant pressure surfaces.*

displaced toward the warmer air.

Cold Core Lows

A cold core low has colder temperatures near its center than those of the surrounding area. A cold core low increases in intensity aloft. The rather weak surface low on the west shore of Hudson Bay (Figure 58) is a major low at 500 MB (Figure 61). Observe that the 500 MB low has been displaced from its surface position toward the colder air. Most troughs (valleys in pressure surfaces) are cold systems. They normally tilt toward the west or northwest with increasing altitude because this is usually the direction of colder air at mid-latitudes.

Fig. 64 *Cross section showing how cold highs decrease in intensity aloft and become cold lows at upper altitudes.*

Warm Core Lows

A warm core low has warmer temperatures near its center than the temperature of the surrounding area. Warm core lows decrease

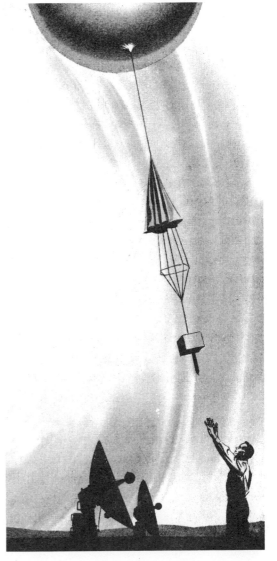

Fig. 63 *Radiosonde equipment being released.*

in intensity aloft and often turn into highs aloft. The surface lows near Panama (Figure 59) underlie a major ridge at 500 MB (Figure 61).

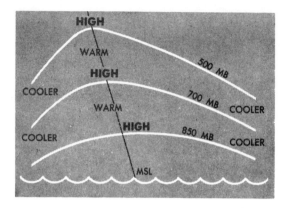

Fig. 65 Cross section showing how warm highs increase in intensity aloft and slope toward warmer air.

Dynamics

While temperature differences between land and water, pole and equator initiate atmospheric motion, they are not the whole story behind pressure distribution. The subject of dynamics of the atmosphere (study of atmospheric motions in terms of the forces involved) is beyond the scope of this book. It is one of the most difficult and complex areas of meteorology. However, the pilot must understand that air in motion in itself can cause pressure changes.

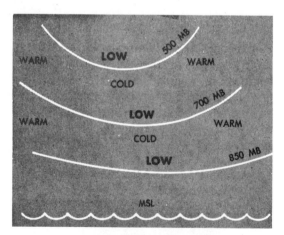

Fig. 66 Cross section showing how cold lows increase in intensity aloft and slope toward colder air.

For instance, major north-south mountain ranges can cause air to pile up. The windward sides of these ranges would then be favorable locations for high pressure areas. Moreover, major troughs are often observed downstream

or on the lee-side of these major ranges. Upper troughs often cause a "bottleneck" to the flow of air and produce a favorable area for surface low pressure formation downstream. Factors such as these contribute to the cold lows and warm highs so often observed.

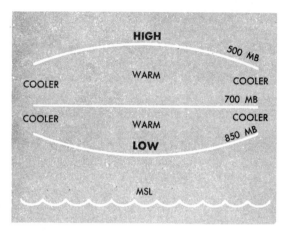

Fig. 67 Cross section showing how warm lows decrease in intensity aloft.

ALTIMETERS AND ALTIMETRY

Knowing the aircraft's altitude is vitally important to the pilot for several reasons. He must be sure that he is flying high enough to clear the highest terrain or obstruction along his intended route; this is especially important when visibility is poor. To keep above mountain peaks, the pilot must note the altitude of the aircraft and elevation of the surrounding terrain at all times. To reduce the potential of a midair collision, the pilot must be sure he is flying the correct altitudes in accordance with air traffic rules (on flights conducted at more than 3,000 ft. above the surface). Often he will fly a certain altitude to take advantage of favorable winds and weather conditions. Also, a knowledge of the altitude is necessary to calculate true airspeeds.

Altitude is vertical distance above some point or level used as a reference. There may be as many kinds of altitude as there are reference levels from which to measure. However, pilots are usually concerned with five types of altitudes:

ABSOLUTE ALTITUDE—The altitude of an aircraft above the surface of the terrain over which it is flying.

INDICATED ALTITUDE—That altitude read directly from the altimeter (uncorrected) after it is set to the current altimeter setting.

PRESSURE ALTITUDE—The altitude read from the altimeter when the altimeter setting window is adjusted to 29.92 (Used for computer solutions for density altitude, true altitude, true airspeed, etc.)

TRUE ALTITUDE—The true height of the aircraft above sea level—the actual altitude. (Often expressed in this manner: "10,900 ft. MSL.") Airport, terrain, and obstacle elevations found on charts and maps are true altitudes.

DENSITY ALTITUDE—This altitude is pressure altitude corrected for nonstandard temperature variations. (An important altitude, since it is directly related to the aircraft's takeoff and climb performance.)

U.S. Standard of Atmosphere Values

Feet	Pressure (in. of mercury)	Temperature (degrees Centigrade)
16,000	16.21	− 17
15,000	16.88	− 15
14,000	17.57	− 13
13,000	18.29	− 11
12,000	19.03	− 9
11,000	19.79	− 7
10,000	20.58	− 5
9000	21.38	− 3
8000	22.22	− 1
7000	23.09	1
6000	23.98	3
5000	24.89	5
4000	25.84	7
3000	26.81	9
2000	27.82	11
1000	28.86	13
Sea Level	29.92	15

THE ALTIMETER

The altimeter measures the height of the aircraft above a given level. Since it is the only instrument that gives altitude information, the altimeter is one of the most important instruments in the aircraft. To use his altimeter effectively, the pilot must thoroughly understand its principle of operation and the effect of barometric pressure and temperature on the altimeter.

Air is more dense at the surface of the earth than aloft. As altitude increases, the atmospheric pressure decreases. This difference in pressure at various levels causes the altimeter to indicate changes in altitude. The pressure altimeter is simply a barometer that measures the pressure of the atmosphere, and presents an altitude indication to the pilot in feet. This indicated altitude is correct, however, only if the sea level barometric pressure is 29.92″ Hg (inches of mercury), sea level free air temperature is + 15° C. (59° F.), and temperature and pressure decrease at a standard rate with increase in altitude. These conditions are requisite to a standard atmosphere, and without appropriate corrections, it is only under standard atmospheric conditions that this type of altimeter is accurate.

ALTIMETER ERRORS

Atmospheric pressure and temperature vary continuously. Rarely is the pressure at sea level 29.92 inches of mercury or the temperature exactly 59° Fahrenheit (standard sea level conditions). If no means were provided for adjusting altimeters to nonstandard pressure, flight could be very hazardous.

On a warm day expanded air is lighter in weight per unit volume than on a cold day, and the pressure levels are raised. For example, the pressure level where the altimeter indicates 10,000 ft. will be HIGHER on a warm day than under standard conditions. On a cold day the reverse is true, and the 10,000-foot level would be LOWER. The adjustment made by the pilot to compensate for nonstandard pressures does not compensate for nonstandard temperatures. Therefore, if terrain or obstacle clearance is a factor in the selection of a cruising altitude, particularly at higher altitudes, remember to anticipate that COLDER-THAN-STANDARD TEMPERATURE will place the aircraft LOWER than the altimeter indicates. (See Figure 69)

Most altimeters are equipped with an altimeter setting window (sometimes referred to as the Kollsman window) which gives the pilot a way to adjust his altimeter for the atmospheric pressure variations discussed previously. FAA regulations provide the following concerning altimeter settings:

The cruising altitude of an aircraft below 18,000 ft. MSL shall be maintained by reference to an altimeter *that is set to the current reported altimeter setting of a station along the route of flight and within 100 nautical miles of the aircraft. If there is no such station, the current reported altimeter setting of an appropriate available station shall be used—and provided further that, in an aircraft having no radio, the altimeter shall be set to the elevation of departure or an appropriate al-*

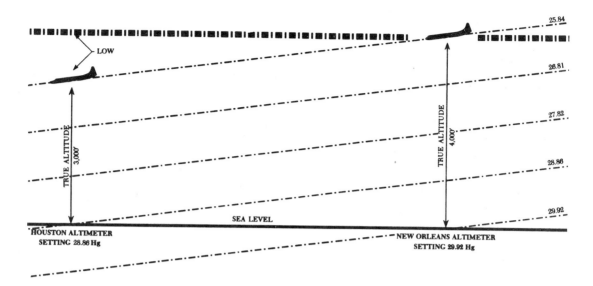

Fig. 68 True altitude decreases when flying into area of low pressure unless the altimeter setting is changed.

timeter setting available before departure.

Many pilots confidently expect that the current altimeter setting will compensate for irregularities in atmospheric pressure at all altitudes. Unfortunately, this is not always true. Remember that the altimeter setting broadcast by ground stations is the *station pressure corrected to mean sea level*. The altimeter setting does not account for distortion at higher levels, particularly the effect of non-standard temperature.

However, it should be pointed out that if each pilot in a given area were to use the same altimeter setting, each altimeter would be equally affected by temperature pressure variation errors, making it possible to maintain desired altitude separation between aircraft.

When flying over high mountainous terrain, remember that certain atmospheric conditions could cause your altimeter to indicate an altitude of 1,000 ft., or more, HIGHER than you actually are. Allow yourself a generous margin of altitude—not only for possible altimeter error, but also for possible downdrafts which are particularly prevalent if high winds are encountered.

As an illustration of the use of the altimeter setting system, we will follow a flight from Love Field, Dallas, Texas, to Abilene Municipal Airport, Abilene, Texas, via the Mineral Wells VOR. Before takeoff from Love Field, Dallas, Texas, the pilot receives a current altimeter setting of 29.85 from the control tower. He applies this setting to the altimeter setting window of his altimeter. He then compares the indication of his altimeter with the known field elevation of 485 ft. If his altimeter is perfectly calibrated, the altimeter should indicate the field elevation of 485 ft. (However, since most altimeters are not perfectly calibrated, an indication of plus or minus 50 ft. is generally considered acceptable. If an altimeter indication is off more than 50 ft. the instrument should be recalibrated by an instrument technician.)

When the pilot is over the Mineral Wells VOR, he makes a position report to the Mineral Wells FAA Flight Service Station. He receives a current altimeter setting of 29.94, which he applies to the altimeter setting window of his altimeter. Before entering the traffic pattern at Abilene Municipal Airport, he receives a new altimeter setting of 29.69 along with other landing instructions from the Abilene tower. Since he desires to fly the traffic pattern at approximately 800 ft. above terrain—the field elevation at Abilene is 1,778 ft.—he maintains an indicated altitude of approximately 2,600 ft. Upon landing, his altimeter should indicate the field elevation at Abilene Municipal (1,778 ft.).

Let's assume that the pilot neglected to adjust his altimeter at Abilene to the current setting. His traffic pattern would have been approximately 250 ft. below the proper traffic pattern altitude, and his altimeter would have indicated approximately 250 ft., more than

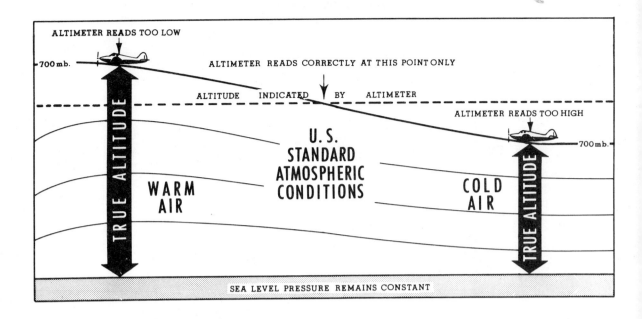

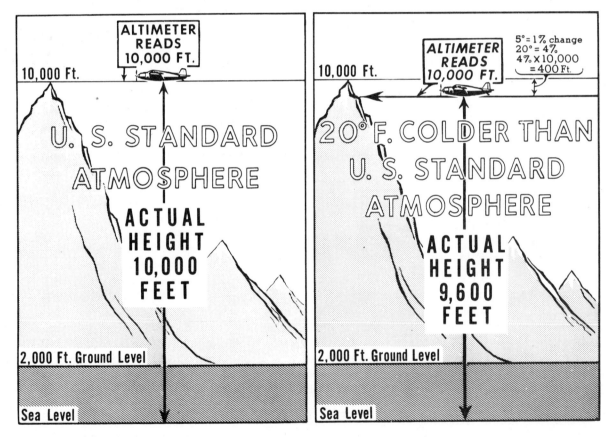

Fig. 69 The altimeter reading is too high when the air is colder, and too low when the air is warmer, than the U.S. Standard Atmosphere.

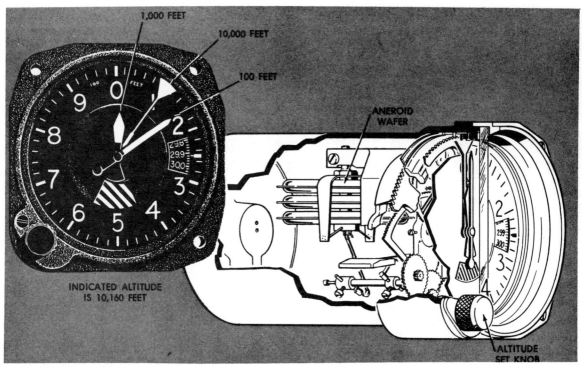

1,000 FEET

10,000 FEET

100 FEET

ANEROID WAFER

INDICATED ALTITUDE
IS 10,160 FEET

ALTITUDE
SET KNOB

Fig. 70 Operation of the altimeter.

the field elevation upon landing.

Actual setting 29.94
Proper setting 29.69
                            ‾‾‾‾‾
                             .25

(1 inch equals)
approximately
1,000 ft.)

.25 x 1,000 ft. = 250 ft.

The importance of properly setting and read-
ing the altimeter cannot be overemphasized.

BASIC WIND THEORY

If it were not for wind, air navigation
would be simple, for a reliable position could
be obtained from only true heading, airspeed,
and elapsed time. It is moving air, sometimes
blowing at speeds of less than 5 knots and at
other times "jet streaming" up to 200 knots,
that complicates navigation. A general understanding of wind theory is, therefore, imperative for the aviator.

Pressure Gradient Force

We have learned that horizontal atmospheric pressure variations do exist. The pressure
equalizing force that tries to drive air from
high to low pressure is called the pressure
gradient force. A pressure gradient is the rate
of change of pressure in the horizontal plane
measured perpendicular to the isobars (or
contours) toward lower pressure. The closer
the isobars (or contours) are together, the

stronger will be the pressure gradient force
and the resulting wind speed (Figure 70).

Coriolis Effect (Force)

When inspecting a constant pressure chart,
it is apparent that the wind is not blowing
from high to low pressure regions, but rather
more along the contours. This phenomenon
is the result of the coriolis force. If the earth
did not rotate, air would flow directly from
high pressure to low pressure. This actually
occurs for small-scale air movements, but
when air moves over a considerable distance,
the air is deflected because of the spinning of
the earth on its axis. In fact, any freely-
moving body is deflected as it moves over the
earth. This deflection is with respect to the
earth. Freely-moving bodies over the earth as
viewed from a point in the solar system, away
from the earth, would appear to follow a
straight line course; but to one stationed on
earth, the path of the freely-moving body
would appear to be a curve. Hence the "force"
is only an apparent force, but the deflection
is a real deflection as far as we on earth
are concerned.

This deflective force is called the coriolis
force, named after G. G. Coriolis, a French
mathematical physicist. The deflection is to
the right of the direction of motion in the
Northern Hemisphere, and to the left of the
direction of motion in the Southern Hemi-

sphere. By convention, all wind directions are designated as the direction from which the wind is blowing.

A wind blowing from Kansas City toward St. Louis would be a west wind, from the west. A west wind without coriolis force becomes an east wind with coriolis force, and so on, in the Northern Hemisphere. Before attempting an explanation of the effects of coriolis force, let us review a few fundamental concepts about the earth.

Since the earth makes one complete rotation each 24 hours, it can be seen in Figure 72 that the linear speed of a point on the

earth's surface decreases with increasing latitude. If a freely-moving body were forced northward from point A to point B, it would enter an area of slower rotating surface of the earth (rotating toward the east) and would show a deflection to the right when viewed from the earth. From space, the body's path would be a straight resultant of the original eastward velocity of the body and the northward thrust.

Imagine twirling a rock tied to a string. As the string is shortened, the rock twirls faster. In the sketch, as the distance of the body from the earth's axis is shortened, an increase

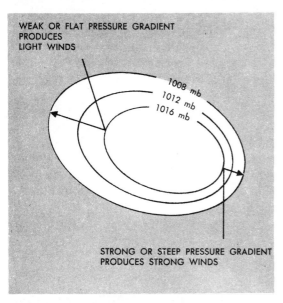

Fig. 71 *Pressure gradient force.*

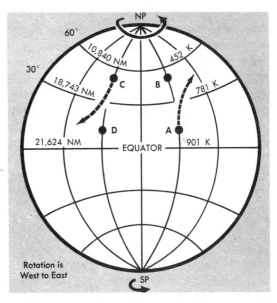

Fig. 72 *Horizontal deflecting force.*

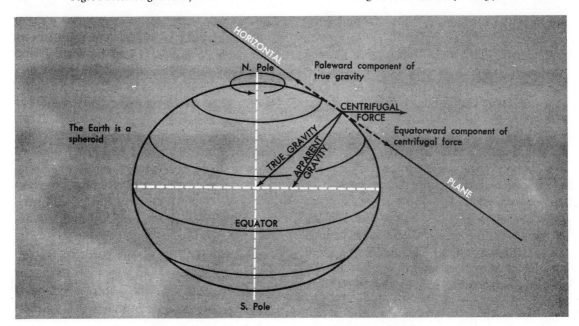

Fig. 73 *The coriolis effect on eastward and westward movement.*

in easterly speed is observed. The ice skater uses this principle when he folds his arms during a spin.

Now consider an unguided body moving from point C toward point D. It will enter an area of faster rotating surface of the earth and will show a deflection to the right again to the observer on the spinning earth. Remember, if the length of the string is increased, the rock will twirl more slowly. In the example of the freely-moving body the distance to the earth's axis has been increased and the body must show a decrease in its original easterly rotating speed. The earth at point D has outrun the body.

It is more difficult to visualize the coriolis effect on eastward and westward freely-moving bodies. Apparent gravity, the force we feel, is composed of true gravity, directed toward the earth's center and the centrifugal force produced by the earth's rotation. Apparent gravity is perpendicular to the earth's surface while true gravity is not. Notice that true gravity has a poleward component. Under the influence of true gravity alone, a body free to move along the earth's surface would come to rest at the pole. For a body at rest with respect to the earth this poleward component of true gravity is balanced by the equatorward component of centrifugal force.

If the body moves eastward, the centrifugal force will increase because the total speed of the body has increased (centrifugal force is proportional to the speed squared) since the body now has a speed around the earth's axis which is the sum of the speed of the earth's surface around the axis plus the speed of the body relative to the earth. The equatorward

component of centrifugal force will also increase. Now the equatorward component of centrifugal force is greater than the poleward component of true gravity (which does not change) and the body is again deflected to the right of its direction of motion.

Lastly, consider coriolis effect on westward moving unsteered bodies. Since the body is moving relative to the earth in a direction opposite that of the earth's rotation, its total speed around the earth's axis is less than that of a body motionless on the earth's surface. Since the speed is less, the total centrifugal force exerted on the body is less, and the equatorward component of the centrifugal force is less. Now the poleward component of true gravity is greater than the equatorward component of centrifugal force, and the body is deflected to the right.

We have seen that regardless of the direction of motion, an unsteered body, such as a parcel of air, will always be deflected to the right of the direction of motion in the Northern Hemisphere. The same conclusion must be reached if one imagines himself riding on the merry-go-round, shown in Figure 74. If a ball is thrown at target "X" without allowing for the rotation of the merry-go-round, the ball will always miss. "X" will move with time to the position shown by the dashed box, and a deflection of the ball to the right will be apparent to the observer on the merry-go-round.

Since the earth is essentially symmetrical with respect to its equator, coriolis deflects to the right in the Northern Hemisphere and to the left of the direction of motion in the Southern Hemisphere. The magnitude of this

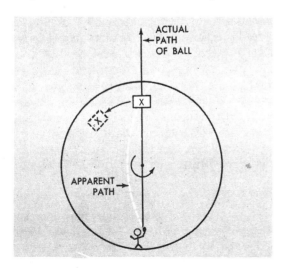

Fig. 74 Merry-go-round example of coriolis effect.

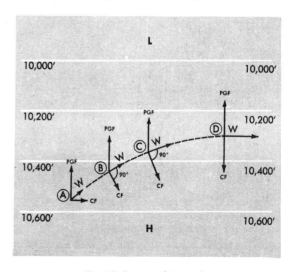

Fig. 75 Geostrophic wind.

deflecting force is a function of several things, two of which are latitude and speed. The coriolis force is zero at the equator and increases toward the poles to become a maximum for any given speed at the poles. It might be noted that the coriolis force is zero for a motionless body at any point on the earth's surface.

GEOSTROPHIC WIND

Considering only pressure gradient and coriolis forces, it is possible to discuss a theoretical wind of great importance to meteorology and navigation. This wind is called the geostrophic wind to signify one resulting from the turning of the earth and pressure gradient forces.

Referring to Figure 75, assume a parcel of air at "A" is subjected to a pressure gradient force (PGF) directed toward the north (as indicated). While the parcel was at rest, the coriolis force (CF) was zero. Now as the parcel initially moves across the contours toward lower pressure, coriolis force increases from zero. The resultant of these two forces is the wind (W). At point "B" the parcel is still moving toward lower pressure and accelerates. Accordingly, the coriolis force increases and the air is deflected more toward the right. This process of accelerating wind speed increasing the deflecting force continues at point "C". At point "D" the pressure gradient force and coriolis force are equal in magnitude and opposite in direction. Without additional accelerating forces, the parcel will move indefinitely in a direction parallel to the contours with the speed it has attained at point "D".

This first wind is blowing parallel to the contours with lower pressure (height) to the left and higher pressure (height) to the right of the direction of motion. Buys Ballot's Law states that in the Northern Hemisphere with one's back to the wind lower pressure is to the left. The reverse is true in the Southern Hemisphere since coriolis deflects to the left there.

This theoretical geostrophic wind demands straight parallel contours and the absence of friction. Happily, above the surface friction layer (generally two to three thousand feet) it is an excellent approximation to the actual wind in the majority of cases. However, at low latitudes (20°N to 20°S) the geostrophic wind is a poor approximation because coriolis force decreases with latitude to zero at the equator, regardless of the wind speed. The

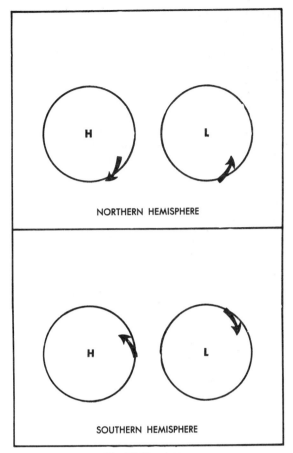

Fig. 76 Circulations.

crosswind component of the geostrophic wind can be computed in flight by the navigator using the radio and pressure altimeters. The geostrophic wind thus forms the basis for pressure differential navigation. It is also used for single heading, minimal, pressure pattern, and optimum flight planning.

Circulations

Consider the accompanying diagrams concerning circulations. Remember that coriolis deflects the wind to the right in the Northern Hemisphere, and to the left in the Southern Hemisphere. The pressure gradient force always acts toward lower pressure. Since the direction of deflection caused by the coriolis force is opposite in the Northern and Southern Hemispheres, we notice clockwise flow about highs in the Northern Hemisphere and about lows in the Southern Hemisphere. We define circulation about lows as always being cyclonic, and the contours are said to have cyclonic curvature. Circulation about highs is said to be anticyclonic, and the contours have anticyclonic curvature.

Gradient Wind

It was pointed out earlier that the geostrophic wind could only be the same as the true wind when the contours are straight and parallel. When the contours are curved, centrifugal force must be considered in addition to the pressure gradient and coriolis forces. The gradient wind is, therefore, simply the geostrophic wind corrected for contour curvature. Sometimes meteorologists speak of the wind at the gradient wind level, meaning not the gradient wind here defined but the wind observed at the top of the friction layer. See Friction, later in this chapter. In Figure 77 the wind at point "A" is geostrophic since the contours are straight and parallel.

The air will strive to continue movement to the west. We treat this tendency of the air to move in its original direction as the effect of centrifugal force. But now the wind is blowing toward higher pressure which cannot continue to any considerable distance as it is countered by the pressure-gradient force. The wind speed will decrease and coriolis and centrifugal forces also will decrease because of their dependence on speed. Pressure-gradient force is now stronger than the combination of the constantly decreasing coriolis and centrifugal forces, and the wind is deflected back toward lower pressure. This battle of forces continues until an equilibrium is reached where the pressure gradient force is equal and opposite to the sum of coriolis force and centrifugal force. This is represented at Point "B" in Figure 77.

Now we again have flow parallel to the contours. But in this equalizing process, what has happened to the wind speed? It has decreased. Therefore, we say that the gradient wind speed about a low is less than the geostrophic speed (for the same contour spacing and latitude).

Now consider the gradient wind about a high, Figure 78. At point "A" we have a geostrophic wind where the contours are straight and parallel. At point "B" we have

the result of the balance of forces where coriolis force equals the sum of centrifugal and pressure-gradient forces, and the flow is again parallel to the contours. During the equalizing process the wind blew across the contours toward lower pressure and thereby increased in speed. This in turn increases the coriolis effect curving the wind toward the right in the diagram. We find that the gradient wind speed about a high is greater than the geostrophic speed (for the same contour spacing and latitude).

The centrifugal force for unit mass is equal to the speed of the body squared, divided by the radius of curvature. The isobars or contours of most highs or anticyclones have a rather large radius of curvature and wide spacing, (slow wind speeds) so that the centrifugal force correction to the geostrophic wind is often unimportant. Moreover, notice that while coriolis force increases with the wind speed, centrifugal force increases with the speed squared. When the speed is large and the radius of curvature small, as in pronounced ridges, coriolis force cannot increase rapidly enough to compensate for centrifugal force, and the wind will not return to flow parallel to the contours.

Friction

Friction with the earth's surface slows the wind speed in the lowest layers of the atmosphere, thereby reducing the coriolis effect. The wind then blows across the isobars toward lower pressure. We have a spiraling inward toward lows and a spiraling outward from highs.

The angle that the wind direction makes with the isobars is usually about 10° over water and about 30° over land. The rougher the terrain, the greater the angle will be. It sometimes reaches 90° over rough land surfaces such as mountainous areas and results in situations where the wind direction shows little relationship to the isobars.

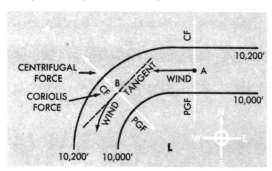

Fig. 77 Gradient wind about a low.

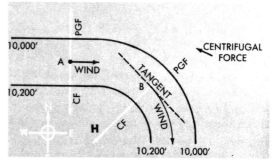

Fig. 78 Gradient wind about a high.

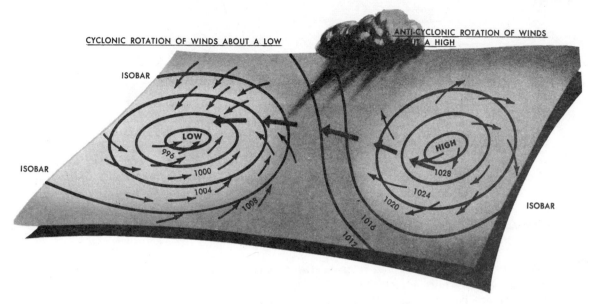

CYCLONIC ROTATION OF WINDS ABOUT A LOW

ANTI-CYCLONIC ROTATION OF WINDS ABOUT A HIGH

ISOBAR

ISOBAR

ISOBAR

Fig. 79 The relation of surface wind to pressure patterns.

The wind flow may be considered frictionless above 1,000 to 2,000 feet over smooth water and above 2,000 to 3,000 feet over smooth land. Over mountainous areas the frictional layer may extend 6,000 feet or more above the range. Above these altitudes the gradient and geostrophic winds are good approximations to the actual wind.

Gustiness

The wind is usually not steady near the surface, especially over irregular terrain, where it moves as a succession of gusts and lulls of variable speed and direction. Eddy currents such as these are caused by friction between air and terrain, and are called gusts or turbulence. This type of turbulence is proportional to the wind speed and the roughness of the terrain.

Gustiness is also produced by locally unequal heating of the earth's surface. Gusts occur when the cooler air of adjacent areas rushes in to replace the rising warm air from heated areas (convection). Trees, lakes, hills and buildings cause variations in surface friction and thereby produce gusts. Gustiness produced by surface friction is usually intensified on a sunny afternoon by this convection from heating at the earth's surface. Gustiness is also quite pronounced on occasions when the wind and pressure systems are such that cool air is advected over warmer terrain causing convection in much the same manner as solar heating.

GENERAL CIRCULATION

Any discussion of the subject of meteorology would be incomplete without mentioning the general circulation of the earth's atmosphere. Yet, the subject of general circulation is one of the most complex of all subjects falling under the heading of meteorology because it involves such a vast area and so many variables. Several theories are found in the literature that attempt an explanation of the mean wind flow of the earth, but none of those so far offered have been universally accepted.

We do have some important facts that can be mentioned here, even though a discussion of all the theories that have been put forth is beyond the scope of this book. We know that on a yearly average the earth receives considerably more energy from the sun in the tropical regions than in the polar regions. Yet, the tropical regions do not continue to get hotter and hotter, nor the polar regions colder. Therefore, something carries heat from the tropics toward the poles. We know, also, that the Northern Hemisphere surface winds follow a relatively constant pattern, with easterly winds generally blowing between the equator and about 30° latitude, westerly winds blowing between 30° and 60° latitude, and easterlies between 60° and 90° latitude. The winds aloft follow a somewhat similar pattern with the midlatitude westerlies extending much farther north.

We observe that season after season, year after year, there tend to be definite belts of high and low pressure along given latitudes. We know that wind tends to blow from high

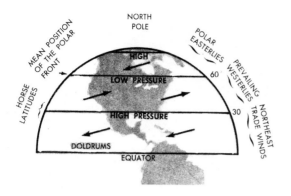

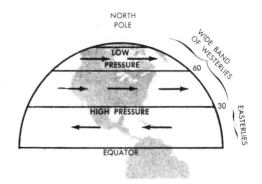

Fig. 80 Diagram showing general surface pressure patterns and resulting wind regimes.

Fig. 81 Diagram showing general pressure patterns at 15,000 feet and resulting wind flow.

pressure toward low pressure and is given direction by the coriolis effect. The winds, therefore, must be the result of the interaction of all the forces acting on the atmosphere. A complete list of these forces would have to include, in addition to the pressure-gradient and coriolis forces, centrifugal force and the various effects that result from friction with the earth's surface, radiation, condensation, and evaporation. So any general circulation pattern that is discussed will not necessarily occur every day in every place, but will simply represent a statistical average over a year's or several years' period of time and must be in agreement with the general pressure patterns.

The mean large-scale pressure and wind patterns have now been established by observations. Published maps of these patterns are available in many weather offices. Figure 80 shows a mean meridional pressure profile at sea level in winter for the Northern Hemisphere and is based on observations, not theories. Notice that low pressure occurs at 60°N even though large high pressure areas occasionally occur at this latitude. Remember this data represents the mean value around all the hemisphere.

Translating the information of Figure 80 onto a view of the Northern Hemisphere and drawing arrows to represent the wind flow pattern at the surface shows how the general circulation for the surface would appear for the Northern Hemisphere.

In general, the pressure pattern aloft at an altitude of about 15,000 feet shows the highest pressure at latitude 30° and lowest pressure in the polar regions, with the largest differential occurring in winter. Figure 81 dis-

plays this pattern in a hemispherical view with the resulting wind arrows drawn. Notice that a wide band of westerly winds occurs, and that there is no band of polar easterlies. Keep in mind that these are mean conditions. Certainly, easterly winds are occasionally found on the polar side of moving low pressure areas aloft in the arctic, particularly when the polar low aloft is displaced some distance from the pole, which often happens.

Thus, we have shown how the winds and pressure patterns must be in agreement, and the observed pressures and observed winds do agree. Why the patterns exist as they do has not been mentioned, since it is in this area that complete agreement is not found. However, the literature does contain several very complete and comprehensive attempts at explaining the "whys" of the general circulation pattern.

Additionally, the mean wind flow gives a good clue to the normal direction of movement of migratory pressure systems. Moving highs and lows, even individual thunderstorms, tend to be "steered" in the direction of the major air flow in which they are imbedded.

Certainly, as more and more information is gathered in this satellite age, new frontiers and explanations of weather systems and the general circulation will be found. Meteorology stands to gain considerably in this space age, since high atmospheric data has been sparse in the past and the solutions to the problems of radiative balance of the earth and the atmosphere must be solved before many weather problems can be solved.

LOCAL WIND EFFECTS

Local pressure and wind systems created by

Fig. 82 Sea breeze, onshore.

mountains, valleys, and water masses are superimposed on the general pressure and wind systems and change the characteristics of the weather of the area. Some local wind effects of interest to pilots are discussed in the paragraphs which follow.

LAND AND SEA BREEZES

Since land masses absorb and radiate heat more rapidly than water masses, the land is warmer than the sea during the day and colder at night. This difference in temperature is most noticeable during the summer months. In coastal areas, this variation of temperature produces a corresponding variation in pressure; during the day, the pressure over the warm land is lower than that over the colder water.

The local pressure and temperature distribution in coastal areas is such that the warm air over the land rises to a higher altitude and then moves horizontally out to sea. To replace this rising warm air at the surface, the colder air over the water moves onto the land (sea breeze). At night, the circulation is reversed so that the air movement at the surface is from land to sea (land breeze). The sea breezes are usually stronger than the land breezes, but they seldom penetrate far inland. Both land and sea breezes are shallow in depth.

The land and water temperature contrast so necessary for land and sea breezes is important for the large scale Asiatic monsoon (seasonal wind). The related illustrations show the January and July mean surface pressure distributions. Notice the low over northern India in July compared to the higher pressure over the cooler Indian Ocean. This pressure distribution causes the onshore, moist, southwest wind of the summer Indian monsoon. The orographic lifting of this air by the Himalaya Mountain Range gives Cherrapunji, Pakistan, an annual rainfall of 35½ feet, falling mostly in the summer months. (Contrast with 35 inches for many areas of our midwest).

During January the extremely cold temperatures over Siberia coupled with the warmer sea temperatures produce strong, dry, cold northwest winds along the Korean coast northward, and northerly winds over India. Observe the great cold surface high over the Asian continent in winter, Figure 84.

MOUNTAIN AND VALLEY WINDS

Sunlit mountain slopes and the air in contact with them are usually warmer than the free air at the same altitude (air farther from the slope) during the day, and they are colder at night. During the day, the air in contact with the slopes becomes lighter than the surrounding air and rises up the slopes. This air movement is called a valley wind because it seems to be flowing up out of the valley as shown in Figure 85.

At night, due to radiation, the air in contact with the mountain slopes becomes colder and denser than the surrounding air farther from the slopes and sinks along the slopes. This air movement is called the mountain breeze because it seems to flow down from the mountains. Mountain breezes, especially in winter, are usually stronger than the valley

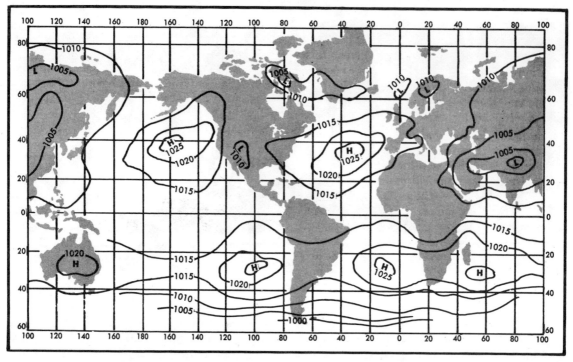

Fig. 83 Prevailing pressure systems of the world, July.

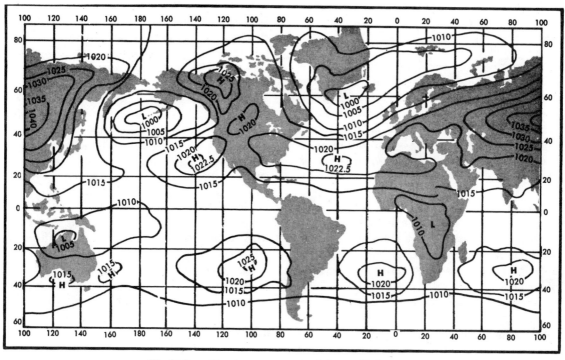

Fig. 84 Prevailing pressure systems of the world, January.

breezes. Notice the respective movements of air in the associated illustration, Figure 86.

KATABATIC WINDS

A katabatic wind is any wind blowing down an incline. If the downslope wind is relatively warm (for the season), it is called a foehn; if the downslope wind is relatively cold, it is called either a fall wind (such as the bora) or a gravity wind (such as the mountain wind).

The name given to the foehn wind found along the eastern slopes of the Rocky Mountains and its associated ranges in the western United States is chinook. The chinook is a warm, dry, downslope (katabatic) wind whose arrival at a weather station located at the base of the mountains may raise the temperature

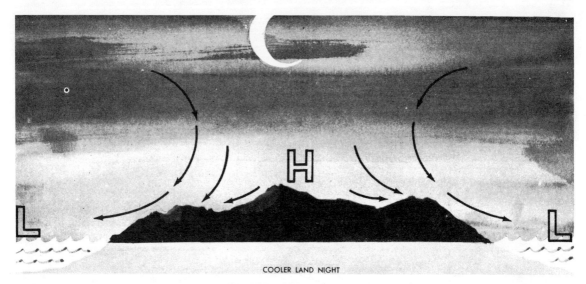

COOLER LAND NIGHT

Fig. 85 Land breeze, offshore.

SUN

Air warming near surface rising along slopes, and generally producing clouds, may obscure mountain peaks.

Fig. 86 Valley wind.

Cooling near surface at night causes denser air to sink along slopes. Air is quite dry and retards formation of fog in the valley.

DRY AIR

Mountain Breezes

Fig. 87 Mountain breeze.

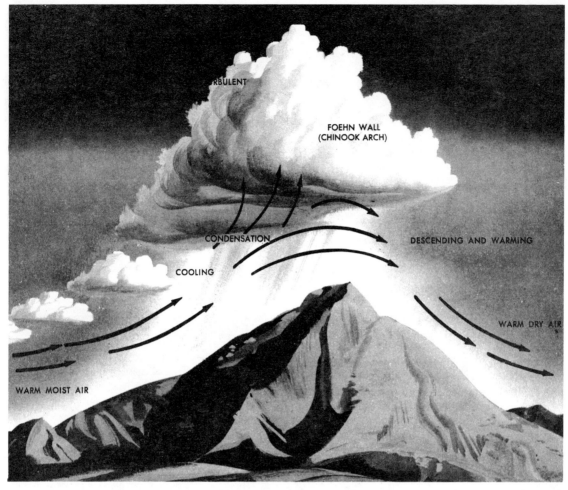

Fig. 88 Sketch showing pattern which produces the warm, dry downslope (Foehn or Chinook) wind. Note the clouds at the mountain peaks.

by as much as 30° Fahrenheit in a few minutes. Figure 87 illustrates the chinook wind situation in which the air rising on the windward side of the mountain is cooled and eventually loses its moisture content through condensation into clouds. The air then continues over the crest of the mountains and descends the leeward slopes into the valley. The downslope motion causes compression of the air and resultant heating, which causes a dry, warm wind literally blowing down from the mountains. The line of clouds that lingers along the crest of the mountains is called a foehn wall (or chinook arch).

The downslope wind which is cold is called a bora. To qualify for the term bora, a wind must be so cold in its source that even after being heated by compression during the descent it arrives in the valley at a lower temperature than the air it is replacing. The best example of the bora is perhaps encountered in winter on the Adriatic coast. The mountain breeze, already discussed, can also be considered a weak kind of a bora, although the term is not usually applied in this case.

The prevailing weather over a location at a given time generally depends on either the character of the prevailing air mass or the interaction of two or more air masses. This chapter will acquaint the pilot with the characteristics of air masses, their formation, and the weather associated with them. The next chapter will deal with the interaction of two or more air masses.

An air mass may be defined as a large body of air with characteristics which are approximately uniform in a horizontal plane. An air mass extends over a large area, usually a thousand miles or more across. The basic properties of an air mass are described in terms of temperature and water vapor content. The weather is usually similar throughout an area covered by the same air mass. However, some modifications do occur because of the local effects of mountains, valleys, and large water masses.

FACTORS WHICH DETERMINE AIR MASS CHARACTERISTICS

The properties of an air mass are determined largely by the surface over which it forms. A body of air that has been nearly stationary or traveling for a long period of time over a region of the earth's surface which has relatively uniform moisture and temperature characteristics develops correspondingly uniform characteristics.

The regions where characteristic types of air masses are formed are called source regions. The source region is the essential determining factor of the initial properties of the air mass. However, air masses that form over a given source region but at different seasons may develop different temperature and moisture characteristics because the characteristics of the source region may vary during the year. The depth to which an air mass takes on the properties of its source region depends upon the length of time it remains there and the stability of the air. Thus, we recognize shallow and deep air masses.

Other factors which determine the eventual characteristics of an air mass are: (1) the characteristics of the surface over which the air mass travels after leaving the source region, and (2) the amount of time that the air mass has been away from its source region.

In order to fulfill the requirements for a good source region, an area must be of uniform surface (either all land or all water), of uniform temperature, and preferably a large area of high pressure where the air has a tendency to stagnate. Many regions of the earth do not fulfill these requirements. For example, most mid-latitude regions are either too variable with respect to winds and temperature, or they have an irregular distribution of land and sea surface. On the other hand, large snow or ice covered polar regions, tropical oceans, and large desert areas adequately fulfill source region requirements.

Classification of Air Masses

An air mass is classified first according to its source region, which may have been either polar or tropical, and these two types are subdivided as to whether they are either continental or maritime.

Air masses which develop in stagnant high-pressure systems of the polar regions are characterized by the low temperatures which they acquire there. They are called cold air masses. Those which develop in the persistent high-pressure systems near 30° latitude are characterized by their high temperatures. They are called warm air masses.

Air masses which originate over large bodies of water usually have a relatively large amount of water vapor in their lower layers. They are called moist or maritime air masses. On the other hand, those which originate over land areas are usually relatively dry, and are called dry or continental air masses.

Accordingly, we may say that there are four basic types of air masses: cold dry, cold moist, warm dry, and warm moist. A pilot will hear the weather forecaster refer to them as continental polar, maritime polar, continental tropical, and maritime tropical. They mean the same thing.

In the Northern Hemisphere, Alaska, Canada, and Siberia are the principal winter source regions for cold, dry air masses while the polar portions of the Atlantic and Pacific Oceans are the main source regions for cold,

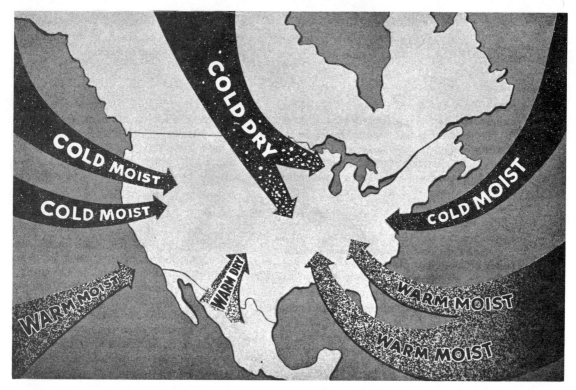

Fig. 89 Trajectories of air masses into North America.

moist air masses.

The tropical regions of the Atlantic and Pacific Oceans are the main source regions for warm moist air masses, and arid regions of Africa, Asia and Australia are the principal source regions for warm, dry air masses. On occasion, the southwestern portion of the United States and the northern portion of Mexico become a source region for warm, dry air masses in North America. Figure 89 shows the trajectories of air masses from their source regions into the North American continent.

In addition to the temperature and moisture content of an air mass, its stability is an important factor in its classification. This is particularly important when it is moving away from its source region. If an air mass is colder than the surface over which it is located, it becomes unstable (i.e., with convective up and down currents) in its lower levels because of the heating which it receives from

Fig. 90 Warm, moist air moving over warmer surface.

Fig. 91 Warm, moist air moving over colder surface.

the warmer surface; that is, it is heated from below. Conversely, when an air mass is warmer than the surface over which it is located, it becomes stable because of the cooling effect of the colder surface; that is, it is cooled from below.

Just as an air mass takes on characteristics dependent on the underlying surface of the region where it forms, so it also tends to have its properties altered by contact with the underlying surface as it moves out of the source region. The degree of modification of an air mass is dependent on the speed at which it travels over the underlying surface, the nature of the surface, and the temperature contrast between the surface and the air mass.

When a cold, dry air mass moves slowly over a warm body of water, both the temperature and humidity of the air mass are increased; also, it becomes less stable. These changes occur in the lowest layers first, of course; when the lowest layers become sufficiently unstable, convective currents tend to spread modifications to progressively higher levels.

From a consideration of the modifying influences of the surface we can anticipate the flying conditions which will generally prevail over a wide area within an air mass. However, in individual localities, there may be special factors which further modify the flying conditions for that locality.

Fig. 92 Heating of land and air by insolation.

Fig. 93 Cooling of land and air by radiation.

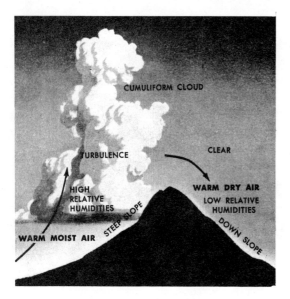

Fig. 94 Orographic lifting resulting in cloud formation, drying of the air and warming of the air on descent downslope.

Figures 90 and 91 illustrate the effects which take place in an air mass due to surface thermal influences. In these cases, the heat content or temperature of the air mass has been changed by the contrasting temperature of the surface over which it is moving.

An air mass over land is often additionally heated during the daytime when the surface is being heated by solar radiation. Similarly, the air temperature near the ground decreases at night when the surface is undergoing radiational cooling (if the sky is clear). (Figures 92 and 93.)

An air mass can also be modified by mechanical influences on its motion which produce lifting, sinking, and mixing actions; these alter the vertical distribution of temperature and water vapor within the air masses. Turbulent mixing occurs when an air mass moves over rough terrain. Sinking motions on the lee sides of mountains increase the temperature and decrease the relative humidity of the air mass. On the other hand, the air mass temperature and relative humidity are altered by passage over a mountain range if clouds and precipitation occur on the windward side. This is shown in Figure 94.

Because of evaporation, the water vapor content of an air mass increases as the air mass moves over a water surface or moist ground, providing the temperature of the water or ground is higher than the dew point of the air mass. Rain falling through an air mass usually raises its humidity.

TYPES AND ACTIVITIES OF AIR MASSES

Thus far, we have been concerned mainly in getting acquainted with how air masses come into being, the different types, and what modifications they undergo as they move over the earth's surface. Now, we are ready to discuss the weather within the different types of air masses. But first, let's briefly review some of the atmospheric activities covered in this and previous chapters. This review will help make the discussion of air mass weather easier to understand.

* The advection of warm air over a colder surface decreases the temperature of the air next to the ground and makes it stable.

* Ascending air currents (convection) are produced by the heat the air receives while in contact with a hot surface, by the air being forced over mountains, or by warm air being forced over colder air.

* The relative humidity of air increases as the air expands and cools in ascending currents. When the lifting and resulting expansion and cooling are extensive enough to produce saturation, clouds and turbulence are the usual result.

* The relative humidity of air decreases as the air contracts and warms in descending air currents. For this reason, clouds generally dissipate in descending air currents.

* Turbulence, vertical air currents, cumuliform clouds, showery precipitation, and good surface visibility (except in showers or dust) are normally associated with unstable air.

* Smooth flying weather, stratiform clouds, and fair to poor surface visibilities are normally associated with stable air.

* Air masses acquire water vapor by evaporation from underlying water masses or from precipitation falling through the air mass (providing the water temperature exceeds the dew point).

* The water vapor content of an air mass is reduced by the formation

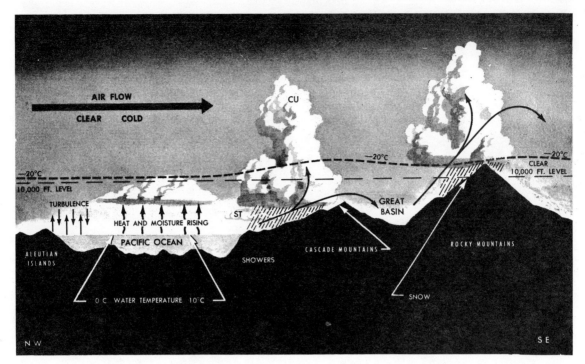

Fig. 95 Winter movement of cold, moist air mass from Gulf of Alaska into Northwest United States.

of clouds and precipitation on the windward side of mountains over which air is flowing. As the air descends on the leeward side, it is heated and consequently dried.

Mountains, valleys, and water masses modify the temperature and humidity of the air mass over a given locality.

There is a daily cyclical variation of temperature in the surface layers of an air mass. Minimum temperatures normally occur near daybreak. The temperature then steadily rises and reaches a maximum value between 1400 and 1600 local time. A steady decrease in temperature (nocturnal cooling) then takes place during the night and early morning.

COLD, MOIST AIR MASSES

The cold, moist air masses which invade the United States arrive from two different source regions — the North Pacific Ocean and the northwestern portion of the North Atlantic Ocean. Cold, moist air masses originating over the North Atlantic appear rather frequently over the northeastern coast of the United States. Those originating over the Pacific Ocean dominate the weather along the Pacific coast of the United States and Canada.

WINTER

The winter weather and flying conditions associated with the cold, moist air invading the Pacific coast vary greatly because of the different trajectories over the ocean which these air masses can have. Those which come from the northwest, that is from the Aleutian Islands and across the Gulf of Alaska, are unstable in the lower layers. Originally very cold, they have gained heat and moisture during the comparatively short trip across the warmer waters of the gulf. Figure 95 shows the typical result. As the air is lifted over the coastal range of mountains, cumuliform clouds are formed. They extend to great heights and their bases will obscure the tops of the mountains.

The cold, moist air masses which approach the Pacific coast more directly from the west probably originated in Siberia, although in some cases they have come from Alaska. Whichever it was, they have had a long overwater trajectory and have picked up considerable heat and moisture. There has been time to distribute it throughout the lower layers so that the air is no longer strongly unstable when it reaches the coast. Stratus and stratocumulus clouds are common throughout the Pacific coast.

Whichever direction they come from, the cold, moist air masses cause extensive precipi-

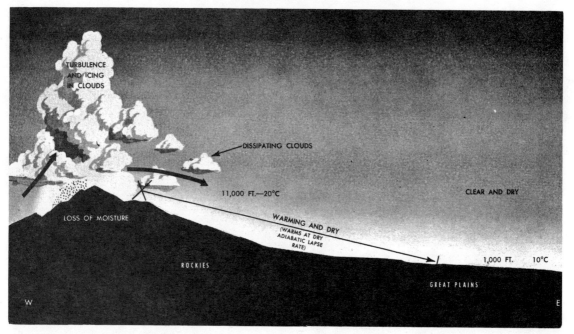

TURBULENCE
AND ICING
IN CLOUDS

DISSIPATING CLOUDS

11,000 FT.—20°C

CLEAR AND DRY

LOSS OF MOISTURE

WARMING AND DRY
(WARMS AT DRY
ADIABATIC LAPSE
RATE)

ROCKIES

1,000 FT.　10°C

GREAT PLAINS

W

E

Fig. 96 Cold, moist air crossing the Rockies.

tation as they move eastward up the western slopes of the mountains. East of the mountains, skies are generally clear, and the air is warm and dry as it comes down the eastern slopes (foehn-like effect).

It is this air mass that causes the Pacific Northwest, including British Columbia, to have more precipitation than other region in North America (especially extreme western Washington in the Olympic Mountains).

Across the country, in the northeastern section of the United States, cold, moist air masses move in from a northeasterly direction. New Englanders use the term "Nor'easter" to describe the weather associated with a strong flow of this air. These air masses are usually colder and more stable than those approaching the west coast from a northwesterly direction; instability is confined to a shallow layer near the surface. Low stratiform clouds form as the air masses move inland.

SUMMER

Although the water temperatures are generally lower than the land temperatures during the summer, the cold, moist air masses are cooled by the cold water belts found along some coast lines. Consequently, they approach the coast in a more stable condition than they otherwise would have.. Fog and low stratus are common along these coasts, some of which are the coasts of California, Peru, Chile, Morroco, southwest Afri-

ca, east Canada, etc. Inland, however, lifting along the western slopes of mountains tends to produce unstable conditions, and cumuliform clouds develop. These are intensified by daytime surface heating. After this air mass has crossed the mountains, it becomes so heated that the relative humidity falls to low values, and clear skies and dry weather prevail.

The summer weather and flying conditions associated with cold, moist air masses over the northeastern United States are similar to those of winter: stratiform clouds or fog.

COLD, DRY AIR MASSES

The cold, dry air masses which invade the United States originate over northern Canada and Alaska; during the winter they also form over the frozen Arctic Ocean.

WINTER

In winter these cold, dry air masses are stable in their source regions. As they move southward toward the United States they are heated by the underlying surface. During the daylight hours the air generally becomes unstable, especially if the sky is clear and there is no snow on the ground. At night the air tends to become stable again.

At times, when these cold, dry air masses move over the warmer waters of the Great Lakes, they acquire a great deal of water vapor and become more unstable. When this happens, cumuliform clouds develop and produce snow flurries to the lee of the lakes, as

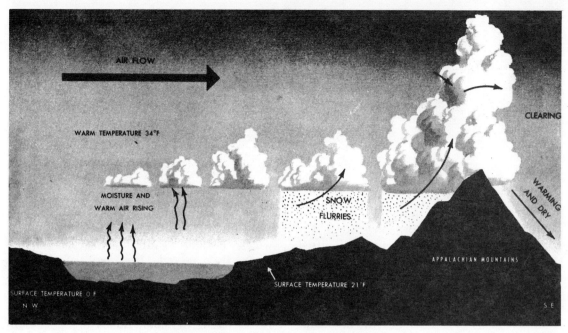

Fig. 97 Cold, dry air crossing the Great Lakes in winter.

shown in Figure 97. In some cases very heavy snowfalls occur close to the lake shores. As the modified air mass moves southeastward, cumuliform clouds build up over the Appalachian Mountains. The skies are usually clear east of the mountains.

SUMMER

Cold, dry air masses have different characteristics in the summer than they do in the winter. Since the thawed land source regions are now warm from solar heating, the air is less stable in the lower layers. The air is cool and dry when it reaches the United States. Cumuliform clouds form during the day when the heating from the sun causes it to become slightly unstable; but the air becomes stable again at night. When these air masses move over the Great Lakes in summer, they are cooled from below, which increases their stability, causing fog. Stratiform clouds are then common over the lakes and to the lee of the lakes. Over the central and southern United States this air is rapidly heated and moistened.

WARM, MOIST AIR MASSES

The source regions of warm, moist air invading middle latitudes are the vast oceanic areas around 15° to 30° latitude. Semipermanent high-pressure systems are located over these regions in all seasons of the year. The one having the greatest influence on the United States is known as the Bermuda high. Depending upon the strength and location of the Bermuda high, warm, moist air masses

originating over the Atlantic Ocean move in over the United States along the southeast coast or through the Gulf of Mexico along the gulf coast. Warm, moist air which originates over the Pacific Ocean is observed only occasionally along the coast of California.

WINTER

Again, because the land is colder than the water, the warm, moist air masses are cooled from below, and as they move inland over the southeastern and Gulf Coast States during the winter, they tend to become stable in the lowest layers. As a whole, however, the warm, moist air mass is very nearly unstable, and in the coastal areas displays a characteristic diurnal variation as illustrated in Figure 87. Fog or low stratiform clouds form at night; they tend to dissipate during the late morning; shortly thereafter, patch cumulus appear, increasing in amount during the afternoon, and disappearing after sundown.

The extent to which the cloudiness and fog spread inland depends on the differences between surface and air temperatures and the strength of onshore winds. When the surface temperatures are cold, the fog and stratiform clouds extend inland for considerable distances if the winds are favorable. On occasion, when the land temperatures are extremely cold, extensive surface temperature inversions develop. Under such conditions, daytime heating usually does not eliminate the surface temperature inversion, and the fog and stratus

0600 LST   1000 L   1200 L   1500 L   1900 L   2100 L

Fig. 98 Diurnal variation in warm, moist air along the Gulf Coast (winter).

may persist for several days.

When the warm, moist air moves far enough over the land surface that it is lifted up the slope of extensive mountain ranges such as the Appalachians, it becomes unstable. The weather and flying conditions are then characterized by heavy cumuliform and stratiform clouds which obscure the mountain tops.

SUMMER

Warm, moist air covers the eastern half of the United States during the greater portion of the summer. Since in the summer the land is normally warmer than the water, particularly during the day, this air mass is heated from below the surface and becomes unstable as it moves inland.

Along the coastal regions, stratiform clouds are common during the early morning hours.

These clouds usually change during the late morning to scattered cumuliform clouds which continue to grow in size and number during the afternoon. Frequently, these cumuliform clouds will develop into extensive thunderstorms by late afternoon. These thunderstorm clouds, however, are generally widely scattered. In mountainous areas, cumuliform clouds and thunderstorms are usually more numerous and intense on the windward side of mountain ranges than on the lee side.

WARM, DRY AIR MASSES

Warm, dry air masses are observed over the western United States primarily in summer, and mainly in their source region which is the Mexico-Great Basin area from where they sometimes spread farther east. This air mass is characterized by high temperatures, low humidities, and sparse cloudiness which,

Fig. 99 Front edge of air mass thundershower. Note reduced visibility. View is from just outside turbulent zone.

when present, is of the cumuliform type and mainly over the mountains. The bases of these cumuliform clouds are exceptionally high for this type of cloud. For example, they may be found as high as 10,000 or 11,000 feet, instead of the more usual heights of 4,000 to 5,000 feet.

The formation of these "low clouds" at such high levels is a graphic indication of the great vertical extent of the effects of surface heating. Flying is often rough at middle and low levels, especially during daylight hours. Occasional dust storms present another significant hazard to flying, since the dust or sand may extend to rather high altitudes and reduce visibilities for extended periods.

Summary

This chapter has described the primary characteristics of the various air masses which influence the weather and flying conditions in the United States. The pilot will encounter many variations of these generalized conditions. To determine the specific conditions for flight within any air mass on a specific day, consult the current weather reports and charts, and discuss the matter with the forecaster.

10 FRONTS

Almost everyone knows that fluids having different density, such as oil and water, do not mix readily. This is also true when two air masses having different characteristics come together. When a cold and a warm air mass come together, the colder, heavier air pushes under the warmer, lighter air. The result is the formation of a front between the two air masses.

Though fronts are strictly associated with weather, the surface of a river, for example, might be called a front. In this case, the front is the surface that separates the water from the air. Such a front is not a fixed surface; there are waves, and the possibility of a rise or fall in the level of the river. This type of front, however, marks the place where the water stops and the air starts. Wherever the water goes, whether into Lake Michigan or the Atlantic Ocean, the front is still there, marking the line where the water stops and the air starts.

Similarly, a weather front may be considered as the line that marks the point where heavy air stops and lighter air starts. The weather front is the surface of the heavier air.

It should be apparent that the contrast between different air masses will be most significant in the lower layers. At some level above the surface, which varies depending on whether the air masses are shallow or deep, the differences between them become small, and the concept of a frontal surface becomes meaningless. As their associated weather is usually confined to the lower troposphere, fronts are seldom recognizable above 15,000 or 20,000 feet, although the temperature change may extend to the tropopause in some cases.

When flying through a front (or when a front moves past a weather observing station), the change from the properties and characteristics of one air mass to those of the other is sometimes quite abrupt (narrow transition zone); at other times, it is very gradual (diffuse transition zone). Fronts tend to be sharpest where there is a tendency for the cold and warm air to blow toward one another (converge).

The polar regions are dominated by cold air masses, the tropics are dominated by warm air masses, and the middle latitudes are regions where cold and warm air masses continually interact with each other — the cold air moving southward and the warm air moving northward — in alternating tongues or waves. The zone which separates these air masses is called the polar front.

The polar front is not stationary. At places a strong flow of cold air pushes southward and replaces the tropical air. At other places it retreats ahead of the advancing warm air. In general we find that while the polar front is advancing southward in one region it is retreating northward in an adjacent region, giving it a wave-like shape, as shown in Figure 88. The polar front, which is observed most frequently in the temperate zone (mid-latitudes), may occasionally move well into the tropics in winter when the cold air masses are dominant. In summer, when the warm air masses are dominant, the polar front may move as far north as 60°N.

Although the polar front is the main zone of discontinuity in each hemisphere, fronts may form between any air masses if the air masses are sufficiently dissimilar.

Fig. 100 Hemispheric view of the polar front. A hypothetical example, showing the semipermanent zone of discontinuity between regions of polar air masses and regions of tropical air masses.

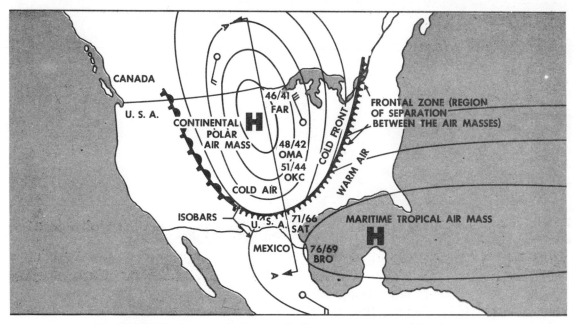

Fig. 101 Weather map showing two distinct air masses separated by a front.

DISCONTINUITIES ACROSS FRONTS

Discontinuities in air mass properties and characteristics, such as temperature, water vapor content, wind, cloud types, and pressure changes are used by forecasters to locate and identify fronts, and to trace their movement. Most of these are shown in various illustrations throughout this chapter. In this section, we shall discuss briefly the discontinuities in temperature and winds which are noticeable. to pilots flying through fronts. Weather hazards are discussed later in the chapter.

Temperature

One of the most easily recognized discontinuities across a front is temperature. At the earth's surface the passage of a front is usually characterized by a noticeable change in temperature. The rate and amount of the change is a partial indication of the intensity of the front. Strong or sharp fronts are accompanied by abrupt and sizeable temperature changes, while weak or diffuse fronts are characterized by gradual or minor changes in temperature.

When flying through a front, you will observe a significant change in temperature, particularly when the course of flight is perpendicular to the front. The change in temperature occurs within a short period of time and/or in a short distance (on the order of 1 to

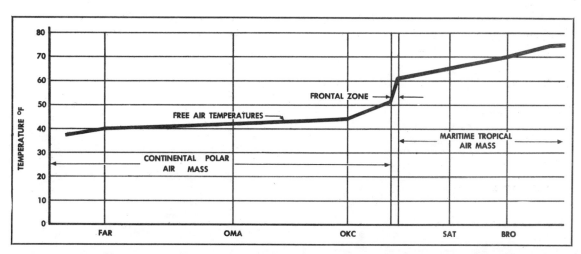

Fig. 102 Gradual uniform change of temperature within air masses shown in Figure 101.

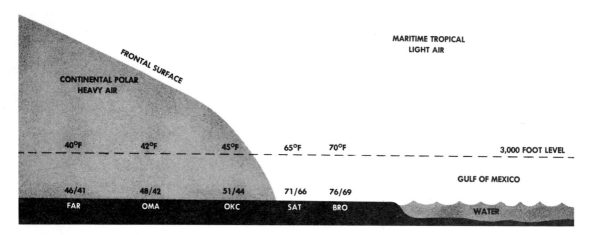

Fig. 103 Vertical cross section along line AA of Figure 101.

20 miles), and is usually more pronounced at lower altitudes than at higher altitudes. The point to remember is that this change, even when gradual, is more pronounced and rapid than the temperature change which may be observed during a flight wholly within one air mass.

The change in temperature is an indication of a change in air mass density across the front. Therefore, it is advisable to obtain a new altimeter setting after crossing a front. If you are flying at a flight level for which a standard setting of 29.92″ is used, this advice does not apply. However, at such an altitude, the temperature change is not likely to be noticeable anyway.

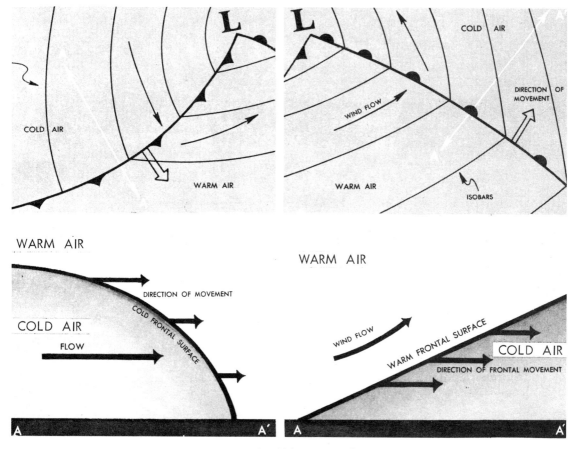

Fig. 104-105 Cold front - warm front.

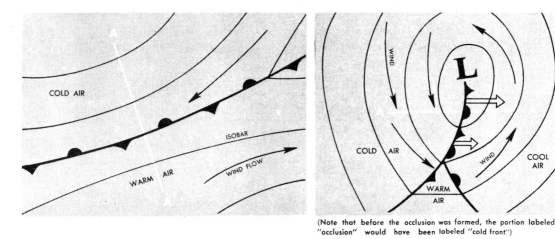

(Note that before the occlusion was formed, the portion labeled "occlusion" would have been labeled "cold front")

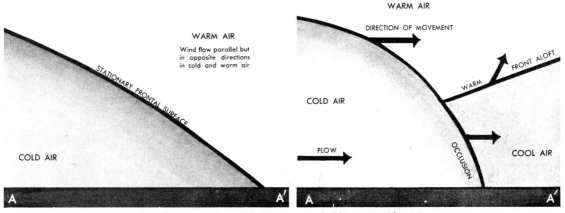

Fig. 106-107 *Stationary front - an occluded front (a cold-type).*

Wind

Near the earth's surface the discontinuity of wind across a front is primarily a matter of a change in the direction. In flying across a front, a simple rule (which applies in the northern hemisphere) is that the wind shift necessitates a change in heading to the right in order to maintain the original ground track.

Wind speed is often very much the same on both sides of a front. In many cases, however, the wind speed increases abruptly after the passage of a cold front and decreases after the passage of a warm front, although it can be the reverse. In general, wind speed is greater in the colder air mass.

CLASSIFICATION OF FRONTS

Fronts are generally classified according to the relative motions of the air masses involved. The four chief classifications are defined here and pictures schematically in the accompanying illustrations.

* A cold front is a front whose motion is such that cold air displaces warm air at the surface (Figure 104).

* A warm front is a front whose motion is such that warm air replaces cold air at the surface (Figure 105).

* A stationary front is a front which has little or no motion (Figure 106).

* The complex front resulting when a surface cold front overtakes a warm front is called an occluded front, or an occlusion (Figure 107).

Frontal Waves and Cyclones

Frontal waves and cyclones (areas of low pressure) are primarily the result of the interaction of two air masses and they usually form on slow-moving cold fronts or stationary fronts.

In the initial condition shown in Figure 108, the winds on both sides of the front are blowing parallel to it (A). Small disturbances to the steady state of this wind, which are often not obvious on the weather map, as well as perhaps uneven local heating and irregular terrain, may start a wave-like bend in the

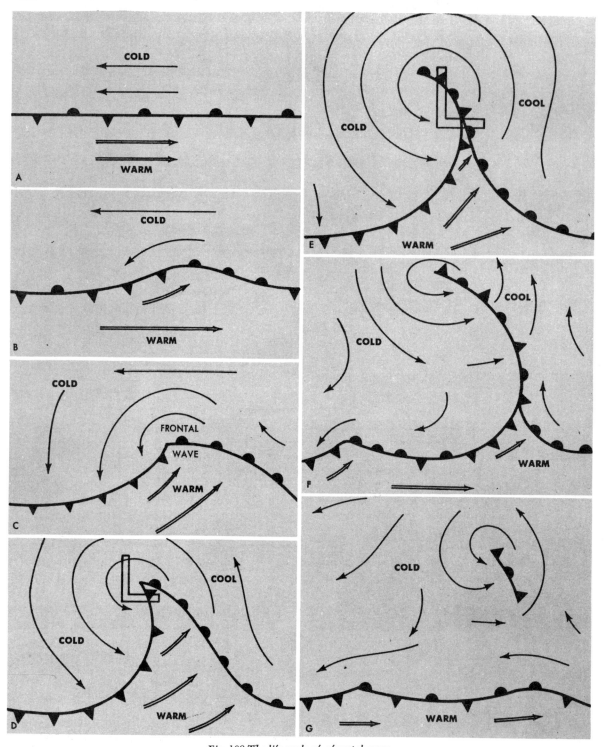

Fig. 108 The life cycle of a frontal wave.

front (B). If this tendency persists and the wave increases in amplitude, a counterclockwise (cyclonic) circulation is set up. One section of the front begins to move as a warm front, while the adjacent section begins to move as a cold front (C). This sort of deformation is called a frontal wave.

The pressure at the peak of the frontal wave falls and a low-pressure center is formed. The cyclonic circulation becomes stronger, and the components of the winds perpendicular to the fronts are now strong enough to move the fronts, with the cold front moving faster than the warm front (D). When the cold front catches up with the warm front, the two of them are said to occlude (close together), and the process or result is called an occlusion (E). This is the time of maximum intensity for the wave cyclone.

As the occlusion continues to extend outward, the cyclonic circulation diminishes in intensity (the low pressure area weakens) and the frontal movement slows down (F). Sometimes a new frontal wave may now begin to form on the long westward trailing portion of the cold front. In the final stage, the two fronts have become a single stationary front again. The low center with its remnant of the occlusion has disappeared (G).

FRONTAL WEATHER

The weather associated with fronts and frontal movements is called frontal weather. It is more complex and variable than air mass weather. The type and intensity of frontal weather is determined largely by such factors as the slope of the front, the water vapor content and stability of the air masses, the speed of the frontal movement, and the relative motion of the air masses at the front. Because of the variability of these factors, the frontal weather may range from a minor wind shift with no clouds or other visible weather activity to severe thunderstorms accompanied by low clouds, poor visibility, hail

and severe turbulence and icing. In addition, the weather associated with one section of a front is frequently quite different from that in other sections of the same front.

A moderate or severe front frequently presents every weather element hazardous to flight; and practically all common weather elements will be found at some position or altitude in the front. The weather can vary greatly from the simple ideal picture, and this may be responsible for the common opinion that no front quite conforms to the weatherman's description of it.

Fronts, like all other weather conditions, have definite life cycles and, accordingly, do not retain their momentary characteristics. The life history of a front may be divided into phases, and within each phase the various weather elements will change in both type and intensity. Hence, in planning a flight a pilot must consider not only the position or intensity of the front at the moment, but also its past history, its tendency to intensify or decay, the effect of terrain upon it, etc. This information can be obtained only from a series of weather maps, with the advice of a forecaster who has been following the frontal activity and is familiar with the characteristics of fronts in the region under consideration.

11 FLYING THE COLD FRONT

Flying the cold front can be a nerve-wracking experience for the pilot untutored in the whys and wherefores of this weather phenomenon. He is liable to encounter rain, sleet, snow, hail, intense icing conditions, and turbulence that often gets violent enough to throw a plane out of control. Also to be found are lightning and occasionally St. Elmo's fire, but these aren't especially dangerous. Everything else about this type front *is* dangerous.

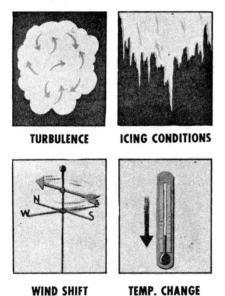

TURBULENCE   **ICING CONDITIONS**

WIND SHIFT   **TEMP. CHANGE**

Fig. 109 Turbulence-icing conditions. Wind shift-temp. change.

There is the hard way to learn to fly cold fronts and that is by flying in them. However, the intent of this chapter is to show you what they can look like, so that you can recognize one; to teach you what is inside the frontal zone, so that you can avoid the danger spots; and to tell you how experienced pilots fly cold fronts, so that you may profit by their knowledge. This knowledge should never blind you to the most important axiom about this phase of flying: APPROACH COLD FRONTS WITH CAUTION!

Cold fronts usually move faster and have a steeper slope than warm fronts. The cold fronts that move very rapidly have very steep slopes in the lower levels and narrow bands of clouds, which are predominantly just ahead of the front. The slower moving cold fronts have less steep slopes and their cloud systems may extend far to the rear of the surface position of the fronts. These characteristics are shown in Figures 113 and 114.

TYPES OF COLD FRONT

Along a cold front, the vertical motion of the air in each air mass, whether upward or downward, will have a definite influence upon the type of cloud formation and weather. The motions may be both ascending, or both descending, or any combination of the two; however, upward motion in the cold air combined

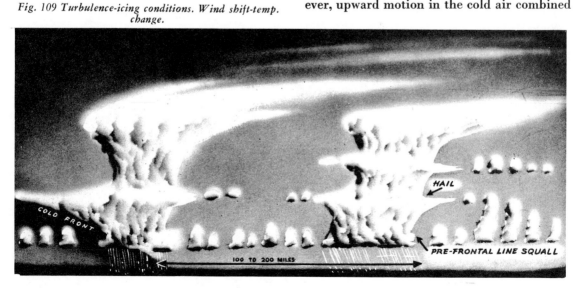

Fig. 110 Cross section through a cold front and prefrontal squall line.

Fig. 111 The cold front - with air descending in both air masses, the front is characterized by absence of clouds and strong winds.

with downward motion in the warm air can occur only when a front encounters the disturbing influence of a mountain barrier. It is obvious in any case that the heavier cold air will tend to underrun the warm air, and the front as a whole will slope toward the center of the cold air mass.

DRY FRONTS SHOULD BE FLOWN HIGH

When the air is descending in both the warm and the cold air masses, condensation cannot occur within them and any clouds present will be dissipated. Such a front is termed a dry, cold front.

The descending air in the forward portion of the cold air mass will, however, bring down with it the high-wind velocity of the air aloft and this type of front is marked by strong gusty winds at the ground. Such fronts produce dust storms over dry terrain and are usually very turbulent.

To a pilot this latter factor is important since turbulent air is a potential hazard to flight. He must never be lulled into a false sense of security simply because no clouds or precipitation are associated with the front. The element of surprise in suddenly hitting extremely turbulent air along a dry front contributes to the possible danger.

Dry fronts should be flown at a high altitude, above 5,000 feet at least. At the first sign of turbulence, the speed of the plane should be reduced to the maneuvering speed (as shown in the airplane flight manual), and the usual sharp "bump" as the frontal surface is crossed should be anticipated. Since the air is descending on both sides of the front, the plane will slowly lose altitude unless held in a gradual climb.

In the cold air mass the region of descending air is restricted to a narrow band behind the front; farther to the rear the air flows horizontally. A secondary front may form between the sinking air, which is being heated by compression, and the horizontally flowing air behind it. The secondary front thus formed may eventually become the predominant front.

The activity of a secondary cold front depends upon the moisture content of the air and the temperature difference between it and the ground. If the ground is the warmer, the forward portion of the cold air will tend to become unstable and the development of squalls is possible. When conditions are favorable, these squalls may develop into well-defined thunderstorms which further complicate the flight problem.

STUDY WEATHER MAPS AND SEE THE FORECASTER

The intensity of ice accretion found in secondary cold fronts of this type will depend upon the moisture content and the degree of instability in the cold air. This is further reason for the pilot to consult the weather map and the forecaster before undertaking flight through the region where a secondary cold front may develop.

Cold fronts with descending air in both air masses usually degenerate, particularly during the day. During the night, however, such fronts tend to become revitalized and the following morning may find them again active. Hence, the history of a front must be followed, and past weather maps as well as the current may and should be used in fore-

Fig. 112 *The cold front - with air descending in the cold mass having an unstable lapse rate, and ascending in the warm mass having a stable lapse rate, squalls will develop in the cold air, and the clouds in the warm sector will be stratified. Secondary fronts generally develop with these conditions.*

casting conditions.

A COMMON TYPE OF COLD FRONT

The weather along a cold front where the warm air is ascending and the cold air sinking depends largely upon the stability of the warm air.

If the warm air is stable, cloud types within it will be stratiform. One or more cloud decks with fairly level bases will extend along the cold front. With continued lifting of the warm air, the upper portion of the cloud mass will reach the ice-crystal level and continuous light precipitation will develop.

When the warm air is stable, the rising motion within it is gentle and steady. Even the added energy coming from the release of latent heat when condensation occurs will not develop any serious degree of turbulence. Therefore, relatively smooth flying is found in the warm air mass. Turbulence will be encountered close to the front.

Icing conditions in the stratiform or nimbostratus clouds along this type of front are usually light, particularly if care is taken in selecting a good flight path. Since the clouds are in layers, flight may be made most of the way between layers where liquid water, except for regions of falling rain, can be avoided.

FLY BELOW THE FREEZING LEVEL

The altitude of the freezing level is important. If the warm air is of tropical origin, the freezing level will be high and the pilot can easily find a flight level where the temperature is above freezing. In the cold air mass, since the air is descending, the formation of cloud will be restricted to local cumulus near the ground. Cloud should be avoided when the temperature is below 0°C.

When the warm air is unstable, or will be made unstable by lifting, the cold front takes on its most intense form, producing severe thunderstorms along the front or in advance of it (prefrontal).

Prefrontal thunderstorms usually reduce the intensity of storms along the cold front. Since the warm air between the prefrontal squall line and the front is usually sinking, the front will be the type first discussed. Such a front can best be crossed by going between the squalls or thunderstorms where, except for moderate turbulence, no serious weather conditions will be encountered.

When the warm air mass is conditionally stable and no prefrontal squall line develops, thunderstorm activity will be confined to the frontal zone. The intensity of the storms may be forecast by studying upper air data or by reviewing the history of the front. The following hints, while not always applicable, will help in estimating the intensity of activity along such fronts.

HINTS ON ESTIMATING COLD FRONTS

The intensity increases as a front moves into warmer areas or where the warm air is more moist.

The greater the wind shear between the warm air and the cold, the more intense the activity.

The steeper the cold front, the more intense the activity, but the shorter the period of frontal passage.

If cold front thunderstorms are active in

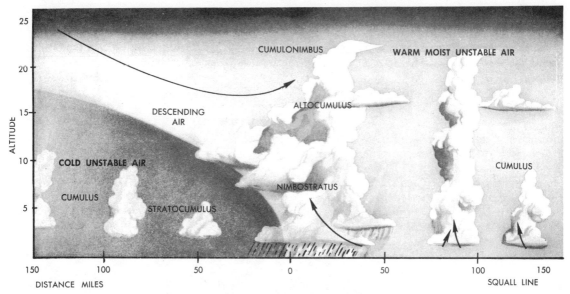

Fig. 113 A fast-moving cold front.

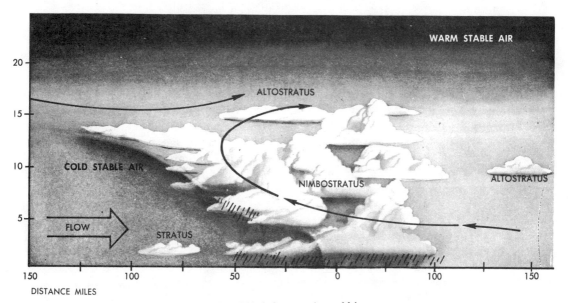

Fig. 114 A slow-moving cold front.

the early morning, the front will usually be severe, with a sharp wind shift and heavy rain. Such a front will not deteriorate during the day, other conditions being equal.

A SPECIAL CASE

Rising air on the cold side of a front and descending air on the warm side are usually only possible when the front crosses a mountain barrier. As a cold front approaches a mountain barrier, rising of the warm air mass is first accentuated. This intensifies the activity within the warm air; therefore, over and to windward of such a barrier, the intensity of thunderstorms is increased. As the cold front reaches the crest of the mountains, the warm

air will flow down the leeward side and clouds within it will dissipate; at the same time, the cold air will be forced upward over the mountain barrier, further lifting the warm air remaining at high levels above it.

EXPECT SHOWERS ON WINDWARD SIDE

Showers may, therefore, continue on the windward side of the mountain. Even after the warm air has all been displaced and forced beyond the mountains, the cold air will continue rising. Unless it is very dry, showers may continue on the windward side of the mountain barrier within the cold air for many hours after the front has passed. This type

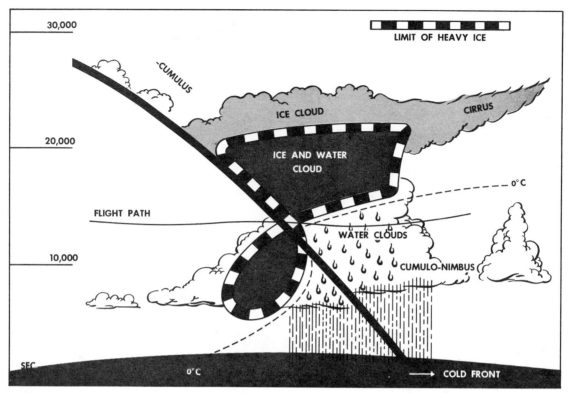

Fig. 115 Icing zones along a cold front.

of weather is common, particularly in winter, on the western slopes of the Allegheny Mountains. Snow and rain squalls frequently persist over Ohio, western Pennsylvania, and the states southwest of this area, for many hours after a front has moved across the mountains.

A cold, stable air mass always endeavors to flow around an obstruction, rather than over it. Therefore, cold fronts are usually retarded upon encountering mountains, though they may move unchecked around either end of a mountain barrier. Backed up against mountains, a front may become nearly stationary for a while. The horizontal extent of the associated weather will then increase and cloud forms will change from unstable to stable types. Thunderstorms and squalls will give way to an extensive overcast with continuous or intermittent precipitation. The cloud's

Fig. 116 The line squall or prefrontal thunderstorm is characterized by the horizontal continuity of form. If you cannot fly over the top—fly just above the roll cloud if terrain permits and be prepared for extreme turbulence.

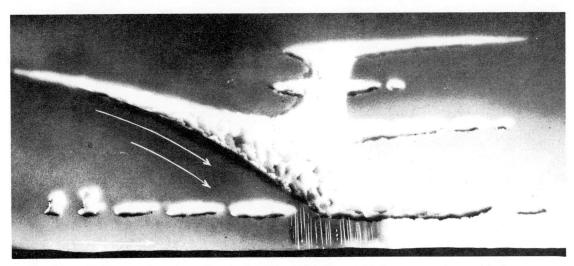

Fig. 117 The cold front - with air descending in the cold mass having a stable lapse rate, and neutral in the warm mass with a conditionally unstable lapse rate, stratified clouds will form in the cold air and the frontal thunderstorms in the warm air mass will be close to the front.

structure will become stratified and the turbulence will decrease.

WHAT TO DO ABOUT IT

In planning a flight through a cold front the following points should be checked:

1. How fast is the front moving? Estimate by wind velocity at the ground, by projecting past movement, or ask the forecaster to compute the movement.

2. What is the vertical motion in the cold air? If descending, cloudiness will be limited to scattered cumulus, or clear with gusty sur-

Fig. 118 Extensive pre-frontal cloud formations (front passed approximately 16 hours later). The winds at the station were south-easterly bringing in moist air from a lower latitude. Towering cumulo-nimbus clouds are in the background with a layer of strato-cumulus topping mountains, including low ragged clouds of bad weather scud.

face winds. If horizontal, the clouds will be stratiform.

3. Are there indications of the formation of a secondary cold front? A region devoid of clouds for some distance behind the front, followed by a broken or overcast sky with low clouds, will indicate such a development. Squalls in the forward portion of the following cloud area are an almost certain indication.

4. What is the stability of the cold air? Stratiform clouds indicate stable air; cumuliform clouds indicate unstable air.

5. What is the altitude of the freezing level? Ask the forecaster, or refer to upper air observations or pilot reports.

6. What is the vertical motion in the warm air? If descending, cloudiness will be patchy and usually cumuliform. If ascending, clouds will be cumulonimbus or nimbostratus. If horizontal, cloudiness will be stratiform.

7. Are there indications that a prefrontal squall line will develop?

8. If flight through the front must be made, where will the front be met? Will you change altitude on passing from one air mass into the other? Will you fly contact, on instruments, or over the top?

A cold front may possess any degree of intensity. An experienced pilot knows this and will be on his guard. He will, in every case, carefully study the factors listed above and will proceed with caution.

The emphasis placed on flying the cold front is not intended to underrate the flight problems that are involved in negotiating the weather conditions to be found in a warm front. So much has been written and there has been so much hangar flying about the difficulties of flying cold fronts that there is a tendency to overlook the fact that warm fronts bring tough flying problems, too.

From the cockpit, warm fronts present two principal hazards.

The first of these is lack of surface visibility and ceiling over a wide area. The second is icing, for there are two distinct sources of icing in the warm front area. A third possible danger exists, too. That of flying full bore into a thunderstorm hidden away in the cloud layers that surround the warm front. While more remote than the ever-present dangers of icing and low visibility, it should not be forgotten.

Warm fronts move at relatively slow speeds and usually have gentle slopes. The speed of the advancing warm air, perpendicular to the front, is greater than that of the retreating cold air. Thus, the warm air not only replaces the cold air at the surface, but also slides up over it along the frontal slope. This active upglide produces a cloud system which, in some instances, may extend as far as 1,000 miles in advance of the surface position of the front.

The type of weather associated with a warm front is governed by the same factors as in a cold front; that is, the stability and vertical motion of the two air masses involved. Therefore, the paths to be followed when flying through a warm front can be tentatively decided upon before take-off, and changes in flight plan will depend upon the pilot's observation of actual conditions. The vertical motion of the air masses is of two types; both air masses ascending, or both descending. Although descending air occasionally occurs along upper warm fronts, it is rare at the surface. Warm front activity of any consequence generally results from ascending air in both air masses.

The activity along a typical warm front develops from the lifting of the warmer air mass to its condensation level as it flows upward over the obstructing wedge of cold air. Eventually, as the upward flow continues, the clouds thus formed pass the ice-crystal level and precipitation occurs. The intensity and extent of warm front activity will depend upon the slope of the front, the stability of the air masses, the height of the condensation, freezing and ice-crystal levels, and the speed with which the warm air flows up the front.

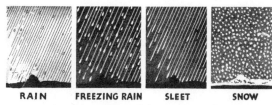

RAIN   FREEZING RAIN   SLEET   SNOW

Fig. 119 You get all four in warm fronts.

The clouds associated with the warm front are predominantly stratiform and appear in the following sequence with the approach of the front: cirrus, cirrostratus, altostratus, and nimbostratus. The cirrus and cirrostratus clouds can be observed hundreds of miles in advance of the front. These clouds thicken rapidly. As their bases gradually lower with the approach of the front, they become altostratus. This is usually between 300 and 500 miles ahead of the front (at the surface). Precipitation may begin to fall from the altostratus.

The amount and type of clouds and precipitation vary with the character of the air masses involved. Three situations are described in the following paragraphs and in Figure 122.

(A) When the over-running warm air is moist and stable, nimbostratus clouds with continuous light precipitation will be found for as much as 300 miles ahead of the front. The bases of the clouds lower rapidly as additional clouds form in the cold air under the frontal surface. These additional clouds which form in the cold air are stratus clouds when the cold air mass is stable, and are stratocumulus clouds when the cold air mass is unstable.

(B) When the over-running air is moist and unstable, cumulus and cumulonimbus clouds (thunderstorms) are frequently imbedded in the nimbostratus and altostratus clouds. In such cases, heavy rainshowers (intense and intermittent) occur along with the continuous light precipitation.

(C) When the over-running warm air is dry, it must ascend to relatively high altitudes before condensation can occur. In these cases, usually only the high and middle clouds will be found.

Visibility is usually good under the cirrus and altostratus clouds. It decreases rapidly in the precipitation area. In addition, when the cold air is stable, an extensive fog area may develop ahead of the front and visibility becomes extremely low in this area.

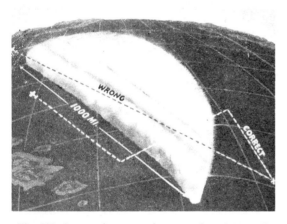

Fig. 120 A warm front may be as much as 1,000 miles in length, so a course parallel to the front and within the weather area would be the height of folly.

Since the cloud system of a warm front is quite extensive, a flight through or along a warm front may require a considerable amount of instrument flying. When the cumulonimbus clouds are embedded in the other

Fig. 121 Thunderstorms lurk in the warm front system.

clouds, they are normally not visible at low and medium altitudes, and sometimes are not distinguishable at high altidues. It is possible to fly into them without any advance warning.

At the surface, the passage of a warm front is characterized by a wind shift, a temperature increase, and an end to the precipitation. There will be a rapid improvement in visibility and a rapid dissipation of the clouds, unless the warm air is moist; in which case clouds, showers, and poor visibility may persist for some time after the frontal passage in the area between the warm and cold fronts (warm sector).

PROBABLE FLIGHT CONDITIONS

Severe icing conditions generally occur in connection with warm fronts since all the elements favorable to intense ice accretion exist. The cloud masses contain large quantities of suspended liquid water, vertical lifting of the air occurs, and the cloud mass is thick and widespread.

In flying through a warm front from the warm to the cold air without changing altitude, the pilot will encounter a rapid drop in temperature. This drop is frequently in excess of 10°C. While the freezing level in the warm air may be 6,000 or 8,000 feet, it may be at ground level a short distance ahead of the front. Therefore, the icing danger in the warm air mass is limited to high levels; but it exists at all levels where clouds or rain occur in the cold air mass. This is particularly true in winter when the warm air is conditionally unstable and nimbostratus clouds prevail along the front. The most intense icing conditions will usually be found where frontal activity is just starting or is increasing. In a region where precipitation has been heavy for some time, the clouds will have lost much of their suspended liquid water.

TEMPERATURE DICTATES FLIGHT PATH

The selection of a flight path under such conditions should be based upon the temperature distribution at the time. If the temperature is below freezing at all levels and the cloud mass is continuous, moderate icing must be expected. The only safe flight path is then at high levels where all clouds may be avoided and those encountered will be very cold and composed mostly of ice crystals.

If when flying from the cold toward the warm air, a fairly close approach can be made to the warm front under the altostratus deck before the nimbostratus cloud mass is

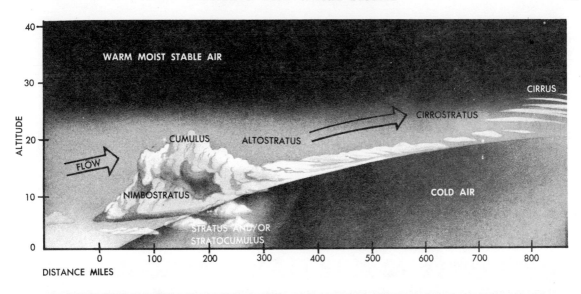

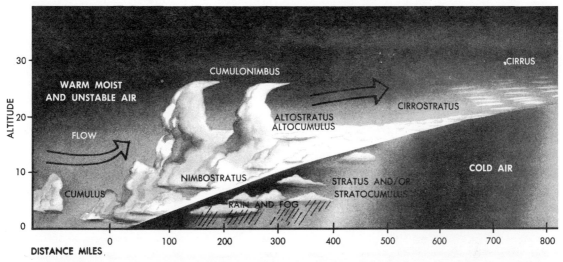

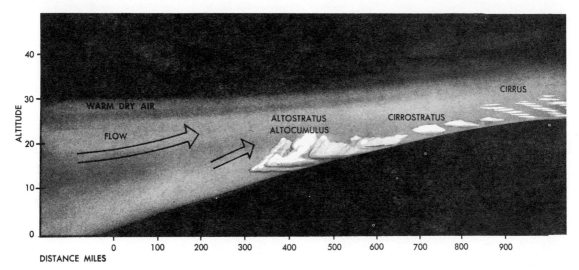

Fig. 122 Three typical warm fronts.

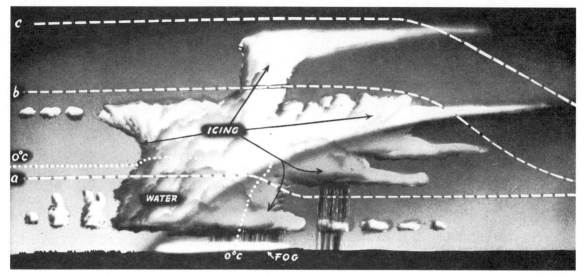

Fig. 123 The warm front - this all-inclusive idealistic cross section illustrates three recommended flight paths to avoid prolonged icing: (a) At an intermediate level on instruments; (b) Over the top, but on instruments at times; (c) Over the top, clearing all clouds.

encountered, it will frequently be possible to fly through clear air to a point where the front can be crossed immediately into above-freezing temperatures in the warm air. Transit through the icing zone can then be made quickly. De-icing equipment should be placed in operation and the front crossed, if possible, at right angles and at an altitude where the temperature in the warm air mass is known to be above freezing.

When flying through the front from the warm towards the cold air, the pilot should make his flight just below the freezing level in the warm air, crossing the front at right angles with the de-icers operating, and maintaining the same altitude until clear air is

found. If clear air is not found shortly after passing through the front, it may be sought by ascending slightly or descending.

When a warm front approaches a mountain barrier, the slope of the front usually steepens to conform to the mountain slope. Steepening of the front intensifies the vertical lift of the warm air, and the clouds will build to a higher altitude. The lower portion of the cold wedge, trapped against the mountain barrier, soon becomes saturated and filled with cloud, with extremely low ceilings prevailing. However, on the leeward side of the mountains, downward flow of the air tends to dissipate the clouds; and if the front can be crossed at that moment the risk of

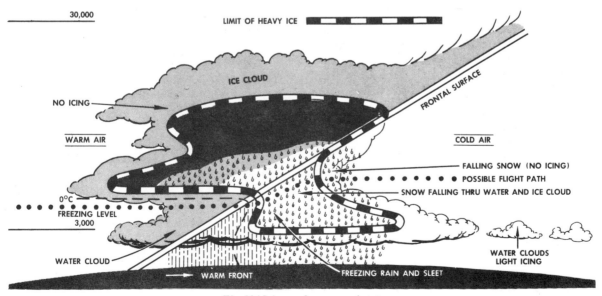

Fig. 124 Icing under a warm front.

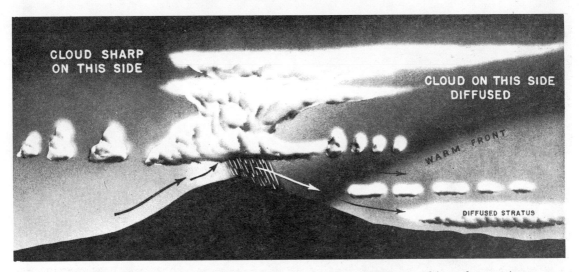

Fig. 125 *On crossing a mountain barrier, the warm front weather is generally separated from the warm front, except stratified clouds persist in the cold air mass.*

icing will be reduced.

WIND VELOCITY MAY DOUBLE

The effect of a mountain barrier is not confined to intensifying the activity of a warm front. Wind velocity increases above the mountains, and at high levels the velocity may double. Thus in crossing a range of mountains against the wind, caution should be exercised. Avoid descending until a definite fix is established and it is certain that the flight path will avoid all obstructions.

Temperature is not a sufficient guide in helping one decide when to let down after crossing mountains, particularly when the warm front is near the crest. Down-slope flow of the air on the lee side of a mountain may warm it sufficiently to make the pilot think he has crossed the front when actually the mountains are still ahead of him.

THE CASE OF STABLE WARM AIR

When the warm air is stable, the clouds along a warm front will be stratiform. This does not mean, however, that layers will not merge. Throughout the nimbostratus along the front, the cloud mass will generally be very deep.

The sequence of clouds encountered when going from the warm to the cold air mass will usually be nimbostratus, altostratus or altocumulus, cirrostratus, and cirrus. Except for the nimbostratus, all are high clouds.

If the warm air mass is relatively dry, it will flow for some distance up to the front before clouds form within it, and the zone of clouds may, therefore, be a 100 miles or more in advance of the surface front. Low clouds may be encountered in the cold air if

rain falls from the warm air above, but they will be stratified and shallow. Icing will be a hazard only at upper levels. This permits contact flight or else flight between layers through the front.

FLIGHT ON TOP USUALLY POSSIBLE

When the warm air is stable, precipitation is generally restricted to the region of the nimbostratus cloud mass. The top of the cloud mass is considerably lower than when the air is unstable, and flight on top is frequently possible.

Fly with caution where stratiform clouds prevail. If any layer of air becomes unstable, dense cloud may quickly be produced which will be an icing hazard in freezing temperatures. Therefore, whenever the temperature is below freezing, flight should be made between layers or on top, entering the clouds only when necessary to change altitude.

PLAN ON LOW CLOUDS

In studying weather conditions, it is important to consider the slope of a warm front and the depth of the cold air. During the summer months, particularly in southern latitudes, the cold air may not be deep enough to lift the warm air above its condensation level. The cloudy area will then be extensive, with low ceilings but little precipitation. During the day the clouds may become broken, but after dark they tend to close in again and lower.

Extensive areas of low cloudiness are usually associated with slow moving warm fronts. Warm fronts usually move more slowly at night. If a flight must be made to a terminal near a warm front, the occurrence of low

Fig. 126 Warm front weather occurring with a conditionally unstable lapse rate. High level showers or thunderstorms well in advance of the front commonly occur with such conditions.

Fig. 127 Warm front weather associated with a convectively unstable warm air mass. This is the heavy precipitation producer. Icing is moderate to severe at all levels above the freezing isotherm.

clouds should be anticipated.

Once a warm front stagnates it seldom regenerates. Ordinarily the front dissipates gradually. It again becomes active only if warmer or less stable air moves over it. The location of the new activity may or may not coincide with the position of the decaying front, depending upon the balance of the various elements involved at the moment.

Remember, the major hazards to flight through a warm front are ice accretion and low ceilings. A satisfactory flight plan can be made only when the limits of the regions occupied by these hazards are known.

When a surface cold front overtakes a warm front, the result is an occluded front. In flight it may be difficult to identify an occlusion because it combines the appearance of both the warm and the cold front. Approached from one direction, it will look like a cold front; from a reciprocal heading it will resemble a warm front.

Mistaking the identity of an occlusion can get you into trouble. The same flight procedures that take you safely through the cold front portion of an occlusion may bring you to grief in the area of warm front weather on the other side. Or, you may choose the correct procedures for the initial warm front conditions only to run head-on into violent activity in the dangerous upper front of the occlusion.

We have already learned in an earlier chapter that fronts frequently have bends, or waves, in them. When these waves exist, one section moves toward colder air as a warm front while an adjacent section moves toward warmer air as a cold front. When the cold front section moves faster than the warm front section, it eventually overtakes the warm front. The warm air mass (warm sector) which was between the fronts is lifted up off the surface by the two colder air masses and shut off from the surface of the earth, hence the term "occluded."

Fig. 128 *From the air, it is hard to tell when you are approaching an occlusion because it combines the appearance of both the warm front and the cold front.*

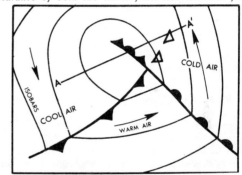

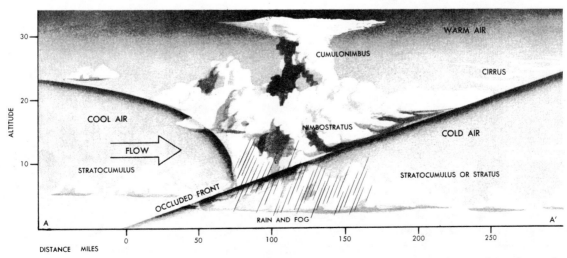

Fig. 129 *A warm-front occlusion. The cross section of the warm front occlusion, occurs at line AA' on the weather map above.*

Fig. 130 A very sharp approach of a leading edge of a warm front occlusion.

There are two types of occlusions — the warm front type and the cold front type.

WARM FRONT OCCLUSION

In the warm-front occlusion, the air ahead of the warm front is colder than the air behind the overtaking cold front. When the cold front overtakes the warm front, both the warm air ahead of the cold front and the cool air behind it slide up over the colder air which is ahead of the warm front. Thus, the cold front itself moves up over the warm front, pushing the warm air ahead of it, and it becomes an upper cold front. At the surface, the situation resembles that found with a warm front (in this case, cool air is replacing cold air), hence the name warm front occlusion. This type of occlusion is not common over the interior of the United States.

The weather associated with warm front occlusions has the characteristics of both warm and cold fronts. The sequence of clouds ahead of the occlusion is similar to the sequence of clouds ahead of a warm front, while the cold front weather occurs near the upper cold front. If either the warm or cool air which is lifted is moist and unstable, showers (and sometimes thunderstorms) may develop. Weather conditions change rapidly in occlusions and are usually most severe during the initial stages of development.

However, as the warm air is lifted to higher and higher altitudes, the weather activity diminishes. Warm front occlusions are found predominantly along the west coasts of continents.

COLD FRONT OCCLUSION

In the cold-front occlusion, the air ahead of the warm front is less cold than the air behind the overtaking cold front. When the cold front overtakes the warm front, both the warm air behind the warm front and the cool air ahead of it are lifted by the colder air coming in behind the cold front. Thus, the warm front itself is lifted by the undercutting cold front, and it becomes an upper warm front. At the surface, the situation resembles that found with the cold front (in this case, cold air is displacing cool air), hence the name, cold front occlusion.

In the occlusion's initial stage of development, the weather and cloud sequence ahead of the occlusion are similar to that associated with warm fronts, while the cloud and weather near the surface position of the front are similar to that associated with cold fronts. As the occlusion develops and the warm air is lifted to higher and higher altitudes, the warm front prefrontal cloud system disappears, and the weather and cloud system become similar to those of a cold front. Cold front occlusions form predominantly over continents or along their east coasts, and are more common than warm-front-type occlusions.

WATCH FOR SAME GOVERNING FACTORS

During the formation of an occlusion, weather conditions along the frontal surfaces will be determined by the same factors which govern warm and cold front weather separately, that is, the stability of the individual air masses, the rapidity with which each air mass is rising or sinking, and the rate at which the various fronts move. There will be other contributing factors, such as variations in terrain, wind velocity, etc.

TIME AND DISTANCE ALL IMPORTANT

In analyzing occlusions, the factors of time and distance are important. Since the occlusion is a progressive development, its character changes with time. As the warm air is lifted to higher levels, the slope of the frontal surfaces diminishes and the weather activity also diminishes. In the following examples each type will be described at the time

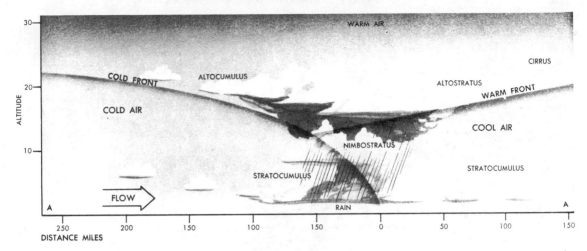

Fig. 131 The cross section of the cold front occlusion shown, occurs at line AA' in the weather map below.

when its activity is most pronounced. This should be borne in mind when actual situations are encountered. The importance of considering the effect of age on an occluded front when planning a flight cannot be stressed too greatly. Age will generally determine whether the air masses are stable or unstable, a condition which further determines the form of the clouds and the type and intensity of precipitation, turbulence, and icing.

As the occlusion develops outward from the center of the low, the most active weather will usually be found at the outer end of the occlusion. The presence of an occlusion normally indicates that the low-pressure system is beginning to stagnate. The deterioration of the storm may proceed rapidly or slowly, depending upon various controlling factors. As the low-pressure area fills, the wind velocity decreases and weather activity along the oc-

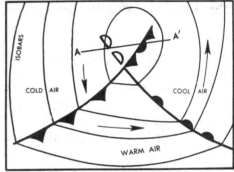

cluded front degenerates.

EXAMPLES OF OCCLUDED SYSTEMS

The following typical examples of occluded systems should be studied carefully. The freezing level has not been indicated on the diagrams because its position will vary with the season and with the history of the individual air masses. With the freezing level in various positions, many different combinations of icing zones are possible. The correct choice of a flight path depends on a thorough

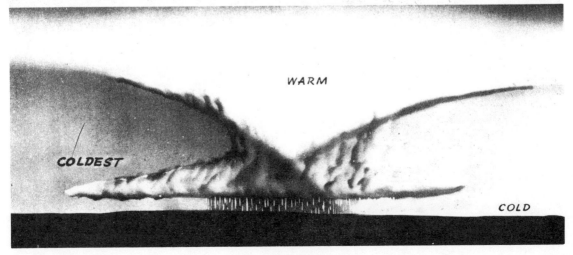

Fig. 132 Occlusion.

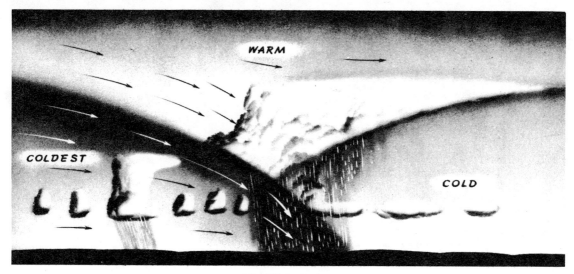

Fig. 133 Occlusion.

understanding of the complicated structure of occluded fronts.

Cold-Type Occlusion

AIR MASS	STABILITY	VERTICAL MOTION
Warm	Stable	Slightly positive
Cold	Stable	Neutral
Coldest	Stable	Neutral

This type of occlusion occurs when a low-pressure area stagnates well away from geographical regions where fronts tend to form. The horizontal motion of the frontal system is usually slow. Winds are light and turbulence is weak.

This cold-type occlusion produces extensive areas of low ceiling, poor visibility, and, if the freezing level is low, moderate icing. Precipitation will be light along the front, with light showers or drizzle elsewhere. If the temperature at the ground is below freezing, freezing drizzle may be present as a definite hazard.

ABOVE 12,000 BEST

The most satisfactory flight path under such conditions is over the top, which is usually above 12,000 feet. Ascent or descent through the altostratus cloud layers along the frontal surfaces will not present a hazard.

When the system has prevailed for more than 24 hours, there is a tendency for the clouds to become stratiform through the frontal zone. Flight between layers may then be made at intermediate altitudes. Such cloud forms will generally be found after continuous precipitation has stopped along that portion of the front which is to be crossed.

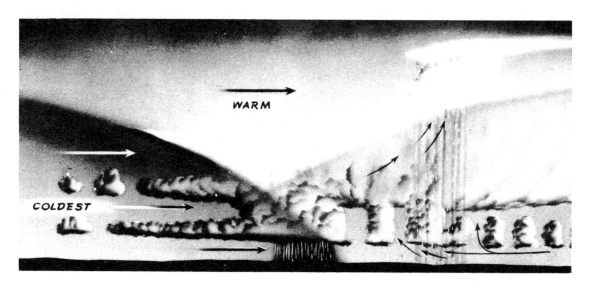

Fig. 134 Occlusion.

Cold-Type Occlusion

AIR MASS	STABILITY	VERTICAL MOTION
Warm	Stable	Negative
Cold	Stable	Neutral
Coldest	Unstable	Negative

This type is common over the western parts of oceans in winter and over eastern parts of continents in summer. The cloud mass in the warm sector is usually dense and unbroken, producing continuous light precipitation. The top of the cloud mass is usually at about 15,000 feet. It is marked by moderate to severe icing conditions, but little turbulence. The low, unbroken ceiling of stratiform clouds will prevail in the cold air mass. Icing will be encountered here along the cold front and the freezing level is below the cloud base.

In the coldest air mass, which is conditionally unstable, the sinking motion near the front provides a zone of broken clouds. Behind the zone where the air is sinking, squalls or showers develop and may assume the characteristics of a secondary cold front.

TRY TO STAY ON TOP OF MOST WEATHER

The most satisfactory flight path is over the top of all weather. A quick descent through the altostratus clouds will not result in any serious icing. Flight at low levels is satisfactory if there are no obstructions and the freezing level is high. When temperatures are low, freezing rain will frequently be encountered in this region.

An alternative flight path is between layers above the tops of the lower clouds. De-icers must be used in flying through the front where the cloud layers merge if the temperature is below freezing.

Cold-Type Occlusion

AIR MASS	STABILITY	VERTICAL MOTION
Warm	Unstable	Up
Cold	Stable	None
Coldest	Unstable	Down

This type of occlusion is common over continents in summer, but is rarely found over oceans. Cumuliform clouds predominate in both the warm and the coldest air masses, with stratiform clouds in the cold air. Icing conditions will be encountered only in the upper portions of cumulonimbus clouds.

Locally low ceilings may occur in the cold air mass, but they are never extensive. At night fog may form under the clouds, usually after the thunderstorms and showers have dissipated.

Secondary fronts may occasionally develop but they never assume the well-defined form which they have when the warm air mass is stable.

Flight through this type of front is best made at low levels below the bases of the thunderstorms. Here the possibility of turbulence is small and the danger of icing negligible.

Cold-Type Occlusion

AIR MASS	STABILITY	VERTICAL MOTION
Warm	Stable	Neutral
Cold	Stable	Neutral
Coldest	Unstable	Vertical

This type of occlusion is frequently mistaken for an upper warm front because of the cumuliform clouds which preceed the passage of the trough aloft and give it the characteristic of an upper warm front.

When the cold air in advance of the front at the ground is unstable, upward motion in it will produce a heavy cloud mass with its upper limit at the warm front surface. The clouds cannot go through this surface for the cold air still remains heavier than the warm air above it.

If the cold air moves over ground warmed by the sun, it may become warm enough to mix with the air above. Clouds can then grow from the lower air mass into the upper one. This "break-through" will usually occur well ahead of the trough. The resulting weather is typical of an upper warm front, that is, instability showers, rising pressure, and a tendency for the clouds to become cumuliform near the front.

WATCH FOR SURPRISES

The "break-through" will usually come in the later afternoon and continue through the night, gradually diminishing in activity. Because previous reports will have indicated an unbroken level cloud top above the warm front, the sudden development of a solid bank of cumulonimbus clouds is most disconcerting.

Warm-Type Occlusion

AIR MASS	STABILITY	VERTICAL MOTION
Warm	Unstable	Positive
Cold	Stable	Neutral
Coldest	Stable	Neutral

This type of occlusion is common over the

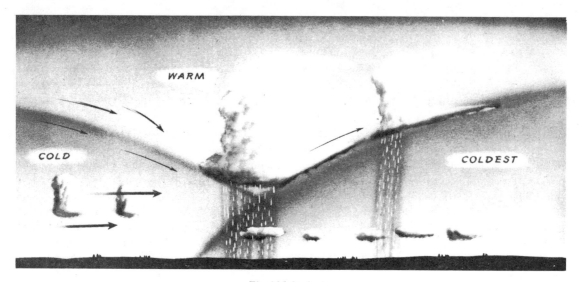

Fig. 135 Occlusion.

interior of continents. It produces high level thunderstorms in summer and may produce light snow in winter. In the latter case, melting snow will cause a narrow band of stratiform clouds at low levels if the lower air is warm. Warm-type occlusions seldom persist when the warm air mass is unstable. Usually the risk of thunderstorms is ended within 36 hours, unless the frontal system is revitalized by a surge of unstable air aloft.

AVOID CUMULONIMBUS CLOUDS

A flight path at low levels above the lowest stratiform clouds is recommended. At higher levels, the pilot should avoid the towering cumulonimbus clouds by flying over or around them.

Warm-Type Occlusion

AIR MASS	STABILITY	VERTICAL MOTION
Warm	Stable	Slightly up
Cold	Unstable	Up
Coldest	Stable	None

This type of occlusion is common in winter over the eastern parts of oceans and the west coasts of continents in middle latitudes. The coldest air ahead of the system will contain only shallow low clouds or no clouds at all. Light rain will fall from the stratiform clouds in the warm air over the warm front. Flight, contact or between layers in the cold air, presents no difficulties.

The occlusion usually moves slowly. Con-

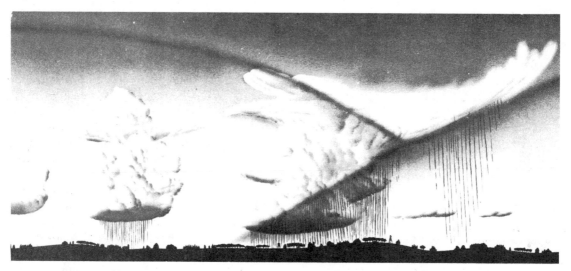

Fig. 136 Occlusion.

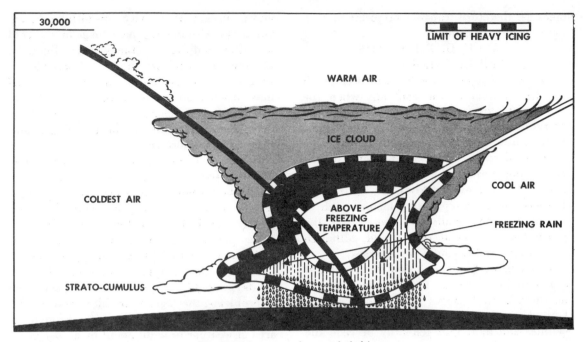

Fig. 137 *Icing zones along occluded front.*

sequently low ceilings and visibilities in the coldest air mass will persist in a given area for a long time. Light rain or intermittent drizzle occurs ahead of the warm front, with showers or squalls in the unstable cold air mass.

If the flow of cold air strengthens, a secondary cold front may develop at point "A"; this front may eventually underrun the warm front and lift all air masses. This develop-

Fig. 138 *Passage of an occlusion direction: West—Precipitation area.*

ment is, therefore, capable of producing an upper front condition.

WATCH OUT FOR THE FRONT ALOFT

The real change of weather will occur with the first arrival of the cold air, when the ceiling will lower suddenly and the amount of rainfall increases as the heavy cloud mass in the unstable cold air comes into play. The cold air deepens rapidly and heavy squalls or thunderstorms take place in the region where the cold air overruns the coldest air. If there are no complicating factors, such as mountains, the squalls will cease after the warm front passes at the ground, although low stratocumulus clouds may persist.

When this type of occlusion moves against a mountain range, as it often does in the northern coastal ranges in the northwestern United States and British Columbia, high ceilings and light rain generally precede the arrival of the cold front aloft. The front passage is followed, however, by heavy snow squalls, low ceilings with severe turbulence, and glaze icing conditions, often with strong shifting winds. These conditions persist until long after the front has passed.

EXTENSIVE FOG AREAS POSSIBLE

Since the instability of the cold air is partly due to heating at the ground, this type of front is most active during the afternoon and evening, with activity diminishing during the night. Ground fog may form in the cold air after midnight, if the wind is light. In conjunction with low ceilings and possible fog in the coldest air ahead of the front, this may produce an extensive fog area. The stable air aloft is characterized by stratiform clouds with a minimum of turbulence and rarely causes more than light icing.

If temperatures near the ground are below freezing, freezing drizzle may occur.

The best flight path is over the top. As a second choice, flight may be made above the low stratified clouds in the coldest air mass. After penetrating the front, the pilot can maintain contact flight by avoiding squalls.

Because the complex nature of occluded fronts presents the pilot with a special set of problems, there are four things you should do if your flight plan takes you through an area of weather brought about by an occluded front.

1. Talk to the weather people about what you can expect to encounter along the route of flight.

2. Study the weather maps carefully.

3. Study the upper air charts for all useful data.

4. Plan your flight carefully and include alternate plans and airports.

Visibility (or visual range) is generally defined as the greatest distance that prominent objects can be seen and identified by unaided normal eyes. As a pilot, you are concerned with the following types of visibilities: horizontal surface (ground) visibility, flight (air to air) visibility, and slant range (pilot's visibility along glide path).

Prevailing horizontal surface visibility is the easiest of the three to measure and is included in surface weather reports. No practical method has yet been devised to determine flight and slant range visibilities, so forecasters depend primarily on weather reports from flyers. Slant range visibility is of vital importance in the approach zone when aircraft, especially the jet type, must land under conditions of low ceiling and/or surface visibility. On many occasions, flight and slant range visibilities are quite different from surface visibility.

Conditions that limit visibility to less than 1 mile present a serious problem only when a landing must be made or when contact flight must be maintained. The main causes of limited visibility are fog, smoke, and dust. Several other weather elements, such as heavy rain or snow, may also reduce the visibility, but they do not present problems as important as the more extensive and persistent conditions of low visibility brought about by fog.

The type and intensity of restrictions to visibility depend largely on the stability of the associated air mass. Stable air is favorable for the formation of fog, low clouds, and light precipitation which restrict visibility. Likewise, haze and smoke are trapped in stable layers of the atmosphere. On the other hand, unstable air produces vertical currents which tend to lift and dissipate fog as well as lift and spread haze and smoke. Blowing dust, blowing snow, and heavy rain showers, which also reduce visibility, are associated primarily with unstable air masses.

In this chapter we will briefly discuss fog, haze, and other phenomena which restrict visibility. Unless specifically stated otherwise, the term visibility in the following discussions refers to horizontal surface visibility.

FOG

Fog is one of the most common and difficult weather hazards encountered in aviation. Since it occurs on the ground, it is primarily a hazard during take-off and landing operations. Fog generally reduces visibility to less than three miles, and on many occasions to zero. Flight visibility is generally good above fog, while slant range visibility is usually near zero in fog. However, fog is not difficult to forecast if good weather reports are available. Forecasting the exact time of formation or dissipation is tricky, and when flight operations depend upon the accuracy of a forecast, the advice of an experienced forecaster should be sought.

Fog is a suspension of minute water droplets in the atmosphere. There is similarity between low clouds and fog. The only distinction between the two is that the base of fog is from the earth's surface upward through 50 feet, and the base of clouds must be at least 51 feet above the ground.

The conditions that are most favorable to the formation of fog are light surface winds, high relative humidities, and an abundance of condensation nuclei. Light winds tend to deepen fog through mixing, although as wind speeds increase beyond certain values (dependent on stability and type of fog) the fog usually dissipates or lifts and becomes low stratus clouds.

Fog is generally more prevalent in coastal areas, where more water vapor is available, than it is inland. Fog is not only more persistent in industrial regions, because of the high concentration of condensation nuclei, but also frequently forms there when the relative humidity is less than 100%. In most areas of the world, fog occurs more frequently during the colder half of the year than the warmer half.

Fog is formed when water vapor in the air condenses, either as a result of cooling the air or of adding water vapor to the air, which in turn leads to the following classification:

1. Fog formed by the cooling process (cooling fogs).

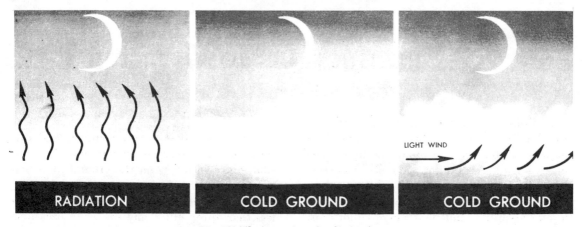

Fig. 139 The formation of radiation fog.

a. Radiation fog — formed by moist air in contact with the earth's surface which is undergoing nocturnal radiational cooling.

b. Advection fog — formed by moist air moving over a colder surface.

c. Upslope fog — formed by moist air which cools because of expansion as it moves up rising terrain.

2. Fog formed by evaporation (evaporation fogs).

a. Frontal fog — formed by evaporation of rain into the colder air mass under a frontal slope.

b. Steam fog — forms when cold air moves over a much warmer water surface.

Fogs which are produced by the cooling process generally occur within an air mass and for this reason are sometimes called air mass fogs. It should be noted that steam fog, which forms primarily within air masses, is produced by a different process: the addition of moisture. The processes that produce cooling fogs also tend to stabilize the lower layers of the atmosphere and thus produce surface inversions.

The classification of fog given here is really a classification of the various processes contributing to fog formation. Actual fogs cannot always be placed in these categories. One of the most common fog situations in the eastern United States is due to a combination of advection and radiation, and steam fog is equally due to advection and evaporation.

Radiation Fog

The atmospheric conditions which are most favorable for the formation of radiation fog are clear skies, light winds, and high relative

Fig. 140 The result of turbulent mixing by light winds.

humidities. These conditions occur most frequently when a land area is under the influence of a high pressure cell.

Radiation fog, sometimes called ground fog, forms on clear, nearly calm nights when the ground loses heat very rapidly. The air in contact with the ground is cooled by conduction, the relative humidity increases, and condensation occurs.

Light wind up to about 5 knots produces a slight mixing of air which spreads the cooling through a deeper layer, as shown in Figure 140. This tends to deepen the fog. If a complete calm exists when all other factors are favorable for the formation of radiation fog, cooling of the air by contact with the ground will be restricted to a shallow layer. Since the cooled air becomes heavier, it will drain into low places and along stream beds. The fog resulting will be shallow. In rough or mountainous terrain it will be restricted to valleys and low areas, but over the plains it may hid the ground over a wide area. Radiation fog seldom forms over snow-covered ground unless the temperature is near freezing.

Most airports are located on low ground, and the flat grass-covered surface favors the formation of ground fog. However, wind movement is also favored by the wide unobstructed space, so that wind tends to slightly delay fog formation. Nevertheless, fog formed in adjacent calm areas can readily drift across the airport. From the air, a pilot may see lights shining brightly through a thin layer of ground fog, and the hazard will pass unnoticed until he levels off for a landing and starts to dip into the top of the fog layer. Forward visibility then drops very suddenly. A pilot should always get a visibility report from the ground whenever there is risk of ground fog.

HOW TO LAND

A safe landing can be made if radiation fog is shallow and somewhat transparent, so that boundary markers, runways, and obstructions can be seen from the air. The procedure to be followed will of course depend upon the pilot's judgment and the type of plane he is flying.

In an emergency the landing must sometimes then be made on instruments. It is best then to "fly" the airplane on to the ground instead of attempting a "stall" landing. The control column should be pushed forward as soon as the wheels touch to keep the air-plane from bouncing or porpoising.

When making a landing with restricted visibility, never look for the ground from the side window. While attention is distracted, the air speed may temporarily be forgotten or the left wing unconsciously lowered, resulting in a stall or ground loop. If a second pilot is available, he should call out the altitude and air speed as the ground is approached. This permits the first pilot to give his undivided attention to holding the plane on an even keel and watching for obstacles and runway markers.

Downwind from large cities smoke will contribute toward lowering the visibility before fog forms. Usually the smoke becomes dense within a few hours after sunset, hanging as a pall at a height of 50 feet or more. When the sky is clear and relative humidity high, water vapor will condense on the smoke particles before the spread between temperature and the dew point is entirely gone, and the visibility will gradually lower. The resulting smoky fog, popularly called "smog," gradually settles to the ground as dew if the air is perfectly calm. The visibility may then temporarily increase before the formation of true radiation fog proceeds.

The increase in visibility after midnight, when a smoke condition has prevailed, should not be taken as indicating an improving trend. If the wind velocity is low and the dew point high, fog is almost certain to follow. If the dew point is lower fog may not form, but the visibility will not improve until after daylight.

On cloudless mornings when smoke is present, the sunlight has a chemical effect upon smoke which causes the visibility to become

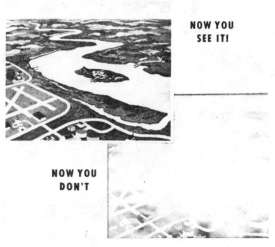

Fig. 141 Now you see it—now you don't.

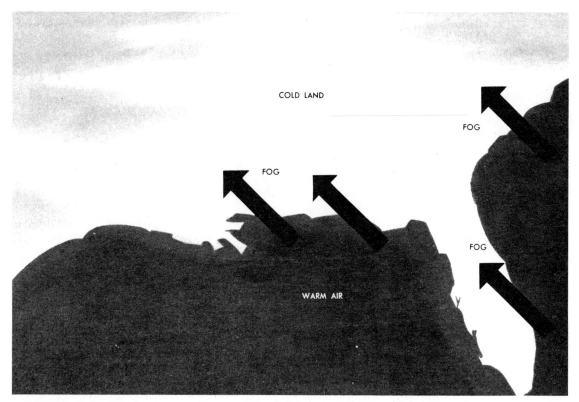

Fig. 142 As the illustration shows, advection fog is common in the Gulf States area during the winter when the relatively warm, moist air from the Gulf moves inland over the colder surface. In summer, the cold water along the east coast of continents frequently produces advection fogs, as warm, moist air moves from land to sea. The sea fog is frequently blown inland during the afternoon by sea breezes.

Fig. 143 Advection fog often forms in coastal areas when cold water from the ocean depths rises to the surface of the sea and produces cold sea-surface temperatures. This type of advection fog, which is shown in the illustration, is common along the California coast in summer. Sea fogs are common over higher latitudes of the oceans in summer as winds from lower latitudes carry moist air over successively colder waters.

Fig. 144 Upslope fog - stronger wind and turbulence lift fog layer as a stratus cloud.

worse for 1/2 to 1 hour after sunrise. When such conditions prevail, the pilot will usually find that the visibility is noticeably worse when he is looking into the sun. If the wind is light, it is usually better to land through smoke with the sun at your back, regardless of the wind direction.

Advection Fog

Advection fog is very common along coastal regions and at sea. It is produced by cooling the lower layers of warm, moist air as the

air moves over a colder surface. Advection fog deepens as the wind speed increases up to about 15 knots. However, when the wind speed is much stronger than 15 knots, the resulting turbulence usually lifts the fog, and stratus clouds form.

Upslope Fog

Upslope fog is formed by the movement of stable air up a sloping land surface. As the illustration indicates, the air rises up the slope and is cooled by expansion. When this cooling

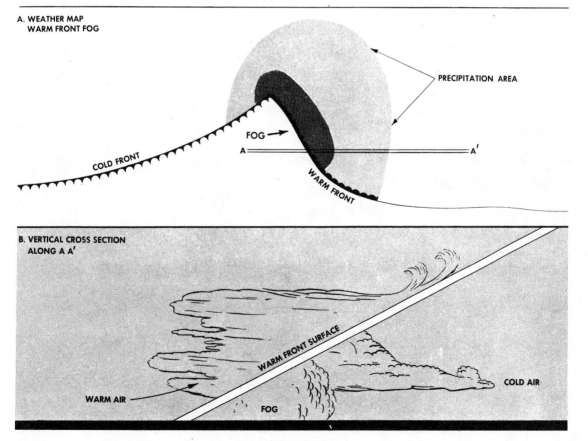

Fig. 145 Warm front fog.

Fig. 146 Steam fog.

is sufficient to produce condensation, fog forms. An upslope wind is necessary for the formation and maintenance of upslope fog. As with other types of fog, when the wind becomes quite strong the fog lifts and becomes low stratus clouds. This type of fog is frequently observed on the high plateaus and eastern slopes of the Rockies when easterly winds from the Mississippi Valley and gulf coast ascend the slopes. It also occurs on the piedmont east of the Appalachians.

Frontal Fog

The most important type of frontal fog forms in the cold air mass under the warm frontal surface. Some of the precipitation from the overrunning warm air evaporates as it falls through and saturates the colder air. Such frontal fogs occur frequently in winter

and are usually associated with warm fronts and occasionally with cold fronts or stationary fronts. Frontal fogs form rapidly and usually cover a wide area.

Steam Fog

Steam fog is produced by the movement of cold air over a warmer water surface. It forms when the evaporation of water vapor into the cold air is intense enough to produce condensation. Steam fog rises from the sea surface like smoke, and therefore is occasionally called sea smoke.

The conditions which produce steam fog — that is, the advection of cold air over a much warmer surface — tend to make the lower layers of the air mass unstable because the air is heated from below. Steam fog is observed over rivers and lakes of the temperate

Fig. 147 Ice fog.

zone in the autumn and over open water areas of the polar regions in winter.

Ice (Crystal) Fog

When the air temperature is below about -25°F, any water vapor in the air condensing into droplets is quickly converted into ice crystals. A suspension of ice crystals in the air at the surface of the earth is called ice fog. Ice fog occurs mostly in the arctic region, and is mainly an artificial fog produced by human activities, occuring locally over settlements and airfields where hydrocarbon fuels are burned. (Burning one pound of a common hydrocarbon fuel produces 1.4 pounds of water.)

When the air temperature is approximately -30°Fahrenheit or lower, ice fog frequently forms very rapidly in the exhaust gases of aircraft, automobile, or other types of combustion engines. When there is little or no wind, it is possible for an aircraft to generate enough ice fog during landing or take-off to cover the runway and a portion of the airfield. Depending on the atmospheric conditions, ice fogs may persist for periods which vary from a few minutes to several days.

There is also a fine arctic mist of ice crystals which persists as a haze over wide expanses of the Arctic basin during winter; it may extend upward through much of the troposphere — a sort of cirrus cloud reaching down to the ground.

OTHER VISIBILITY RESTRICTIONS
Haze and Smoke

When the fine dust, salt particles, or other impurties which are normally highly dispersed in the atmosphere, are trapped and concentrated in a limited layer, the resulting restriction to visibility is called haze. When we view distant objects against a dark background through haze, they appear to be veiled in pale blue, pale yellow, or white, depending on the density, illumination, and nature of the particles. The density of haze near the ground increases as the stability of the air increases. Surface-based haze layers often extend to altitudes of about 15,000 feet through convectional mixing, with a sharp boundary at the top called a haze line or dust horizon when seen from above.

The greatest restriction to visibility in haze usually occurs when looking in the direction of the sun; then the visibility is frequently zero. In fact, it is often hazardous to land an aircraft into the sun when haze conditions exist. Visibility in haze is best in the vertical, particularly looking down.

Smoke usually restricts visibility when it is trapped under an inversion in stable air and is usually concentrated on the downwind side of industrial areas. The surface, slant, and flight visibilities in smoke are similar to those in haze.

Smoke from forest fires is frequently

AIRPORT CLEAR

Fog formed in industrial areas tends to be more persistant than other types of fog.

SMOKE AIDS FOG FORMATION

AIRPORT

FOGGED

IN

Fig. 148 Smog.

transported great distances at high levels. In such cases, very poor flight and slant range visibilities in dense smoke at flight altitudes may be encountered even though the lower levels are free from smoke.

Since smoke particles are nuclei upon which water vapor condenses, and since the conditions leading to radiation fog produce stable air near the ground, smoke and fog frequently occur together in industrial areas. This mixture is sometimes called smog.

Blowing Snow, Dust, and Sand

Blowing dust is observed in semiarid regions when the air is unstable and the winds are relatively strong. The strong winds and vertical currents may spread the dust over wide areas and lift it to great heights. Surface, flight, and slant range visibilities are reduced to very low values in blowing dust. Sand storms are more local and occur where loose sand is found in desert regions. The blowing sand is seldom lifted above 50 feet.

Blowing snow reaching a few feet above the ground observed over snow-covered regions when the wind is very strong can be as troublesome as ground fog.

Precipitation

Rain, except in brief, heavy showers, rarely reduces the surface visibility to less than one mile. However, when rain streams over the windshield of an aircraft, it will greatly reduce the visibility to the outside. This is often called cockpit visibility.

Drizzle and snow usually reduce the visibility to a greater extent than does rain. Drizzle, which is common in stable air, is generally accompanied by fog, haze, or smoke.

The visibility is frequently reduced to zero in heavy snowfall.

Clouds

Flight and slant range visibilities vary from about one-half mile in light or thin clouds to zero in dense clouds. Since there is little or no slant visibility until the pilot penetrates the base of the cloud, low ceilings present visibility problems during landing operations.

Surface-Based and Total Obscurations

When the sky or clouds are partially or totally hidden from a ground observer by precipitation or by other restrictions to visibility (smoke, fog, haze, and the like) whose bases are on the ground, the sky is said to be obscured.

When the sky or clouds are totally obscured, the reported ceiling is the vertical visibility from the ground. For example, if all the sky is hidden by fog and the vertical visibility is 500 feet, an obscuration ceiling of 500 feet is reported. This obscuration ceiling of 500 feet is quite different from a cloud ceiling of 500 feet. In the latter case, the pilot can normally expect to see the ground and the runway on the glide path when the aircraft penetrates the cloud base at 500 feet above the ground. This situation is sketched below.

However, in the case of the obscuration ceiling, the obscuring phenomenon (precipitation, fog, haze, smoke, or the like) reaches to the surface and the pilot normally will not be able to see the runway or approach lights during an approach, even after penetrating the level of the reported obscuration ceiling. Notice this situation in Figure 150 below. The pilot will be able to see the ground directly beneath his aircraft when he passes through the altitude of the reported vertical visibility — 500 feet and, of course, will often

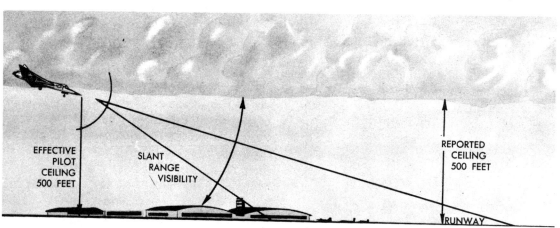

Fig. 149 Cloud ceiling of 500 ft. with no restriction to visibility below clouds.

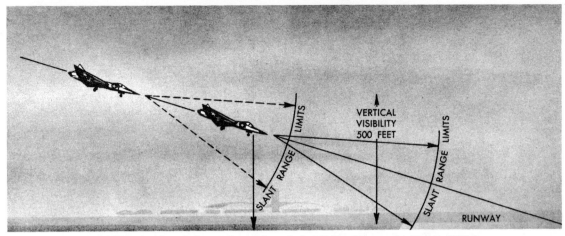

Fig. 150 Surface based obscuration with vertical visibility (obscured ceiling) of 500 ft.

make visual contact with the runway only after reaching an altitude considerably lower than the reported vertical visibility.

Of more concern than seeing the ground directly beneath his aircraft is seeing some distance ahead of the aircraft and, more specifically, the runway or approach lights.

TELETYPE REPORT EXAMPLES
Total Obscurations
1. W1 X 1/16 F

The symbols mean respectively:

W — indefinite ceiling; in the case the obscuration is fog.

1 — vertical visibility estimated to nearest 100 feet — thus it is 100 feet in the example.

X — total obscuration. The entire sky dome and clouds are obscured from the observer.

1/16 — horizontal prevailing visibility is 1/16 statute miles.

F — symbol for fog.

This would be read: "indefinite ceiling one hundred feet obscured, visibility 1/16 of a mile in fog."

2. W5X½S

The symbols mean respectively:

W — indefinite ceiling; in this case the sky dome or clouds are obscured by falling precipitation, which is snow.

5 — vertical visibility estimated in hundreds of feet — thus it is 500 feet in the example.

X — total obscuration.

½ — horizontal prevailing visibility in statute miles and fractions thereof. In the example it is ½ of a mile.

S — snow. In the example the S means snow of moderate intensity.

Heavy snow would be shown as S+.

Light snow would be shown as S−.

Very light snow would be shown as S--.

This would be read: "indefinite ceiling five hundred feet obscured, ½ of a mile visibility in moderate snow."

Partial Obscurations

When 1/10 through 9/10 of the sky dome or clouds is visible through the obscuring phenomenon at the point of observation, the obscuration is reported as partial. Since the ground observer can see through the obscuration, or a portion of the sky dome is not hidden by the obscuring phenomenon, vertical visibility is not reported for partial obscurations as it is for a total obscuration. However, when clouds are visible with a partial obscuration, their heights and amounts are reported.

Partial obscurations present the same landing operation problems as the total obscuration. The pilot still has no idea of the slant range visibility or the altitude at which he will first see approach lights or runway.

The amount (in tenths) of the sky or clouds obscured by a partial obscuration is included in the remarks section of weather reports. Although this may help to clarify the reported conditions in many cases, it still does not provide an idea of the slant range visibility.

TELETYPE REPORT EXAMPLE
Partial Obscurations

−XM5 ⊕ 3FHK 186/64/62C/003 F6

The symbols mean respectively:

−X — partial obscuration, or that part of the sky dome is hidden by surface-based phenomena.

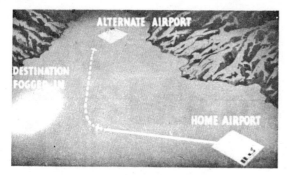

Fig. 151 Use it!

M — indicates the method by which the base of the visible clouds was determined and generally refers to the use of a ceiling light, ceilometer, or cloud base-height measuring instruments.

5 — base of the visible clouds in hundreds of feet. In the example it is 500 feet.

(I) — broken sky condition, or the portion of the sky obscured by the obscuring phenomena plus the portion covered with clouds is 6/10 through 9/10.

3 — horizontal prevailing visibility in statute miles (see below).

F — fog

H — haze

K — smoke

186 — sea level pressure of 1018.6 millibars.

64 — temperature in degrees Fahrenheit

62 — dew point temperature in degrees Fahrenheit

C — surface winds are calm

003 — altimeter setting in inches of mercury; thus, 30.03".

F6 — indicates that 6/10 of the sky dome is obscured by the obscuring phenomena, or, fog 6/10.

HAVE AN ALTERNATE AIRPORT

Because fog and other visibility limiting weather phenomena are hazardous chiefly because they obscure obstructions around airports, always have an alternate airport in mind if there is a chance your destination might be fogged in. It rarely pays to circle over an airport waiting for conditions to improve; you will simply exhaust your fuel supply. The surest safety device in overcoming poor visibility is enough gas to get you to an alternate airport.

Turbulent air is rough on the pilot, passengers, and the airplane. It matters little how much flying time you have or how knowledgeable you are about the causative factors of atmospheric turbulence — you're going to experience the ups and downs of this phenomenon sooner or later.

However, to make it as easy as possible on the three objects mentioned above, adequate preparation of all three is imperative.

Atmospheric turbulence is caused by random fluctuations of wind flow which are instantaneous and irregular — so much so that to accurately describe the random fluctuations by other than statistical analysis is almost impossible. The motion of turbulent air is made up of a series of gusts. A gust is a momentary surge of fast-moving air. It may come from any direction of the compass, or it may come upward, downward, or at an angle. Some gusts have very sharp boundaries in space so that you fly into them with no warning whatsoever. Others are not so sharply bounded, so that you hit them more gradually. How fast the gust is going is not so important as how suddenly you hit it. Of course, a heavily loaded airplane is in greater danger than the same plane loaded lightly.

CAUSATIVE FACTORS

In general, turbulence can be approached from the standpoint of causative factors and described by a subjective scale of degrees of intensity. For the purposes of this discussion we shall divide turbulence according to the following causative factors:

1. Thermal — caused by localized vertical convective currents due to surface heating or unstable lapse rates, and cold air moving over warmer ground.

2. Mechanical — resulting from wind flowing over irregular terrain.

3. Frontal — resulting from the local lifting of warm air by cold air masses, or the abrupt wind shift (shear) associated with most cold fronts.

4. Large Scale Wind Shear — marked gradient in wind speed and/or direction due to general variations in the temperature and pressure fields aloft. Two or more of the above causative factors often work together. In ad-

dition, turbulence is produced by "man-made" phenomena, such as the wake of aircraft.

DEGREES OF TURBULENCE DEFINED

In order to clarify the discussion of turbulence, the following descriptions are given for the various degrees of intensity:

LIGHT TURBULENCE

Light turbulence is defined as a nearly ambient condition of turbulence over extensive areas and at any altitude. The more intense turbulence in this case is experienced in small cumuliform clouds. It is also found at low levels over rough terrain with surface wind speed less than 25 knots. It is experienced at low levels over unequally heated land areas during the period of maximum heating and at night over warm water areas.

MODERATE TURBULENCE

Moderate turbulence is experienced in relation to:

* The mountain wave when the strongest winds at mountain top level perpendicular to the ridge line are 20 to 50 knots or more. Moderate turbulence is *frequently* found from the surface to 10,000 feet above the tropopause and as much as 300 miles leeward of mountains or within cirrus clouds associated with the wave.

* The mountain wave when the strongest winds at mountain top level perpendicular to the ridge line are 25 to 50 knots. Moderate turbulence is *frequently* found between the surface and the tropopause from the ridge line of mountains to 150 miles leeward, or within cirrus clouds associated with the wave.

* Thunderstorms, and is *frequently* found in, around, and above dissipating thunderstorms, or within the cirrus tops.

* The jet stream, and is *frequently* found within a layer between the height of the jet core and 5000 feet below the core of the jet, and from the core to 250 miles toward the cyclonic (cold) side of the core, or within cirrus clouds associated with the jet.

* Cumuliform clouds, and is *usually* found within thick or towering cumulus.

* Strong surface winds, and is *usually* found near the ground when surface winds exceed 25 knots.

Fig. 152 Strength of convective currents vary according to ground conditions.

* Upper trough, cold low, or front aloft, and is *frequently* found where vertical wind shear exceeds 6 knots per 1000 feet or horizontal wind shear exceeds 40 knots per 150 miles.

* Unstable atmosphere, and is *frequently* found at low levels where the atmosphere is unstable but moisture is insufficient for thunderstorms or towering cumulus to form.

SEVERE TURBULENCE

Severe turbulence is defined in relation to:

* The mountain wave when the strongest winds at mountain top level perpendicular to the ridge line are 50 knots or more. Severe turbulence is *usually* found from the surface to the tropopause, and from the ridge line to 150 miles leeward.

* The mountain wave when the strongest winds at mountain top level perpendicular to the ridge line are 20 to 50 knots. Severe turbulence will *usually* be found leeward of mountains up to 50 miles downstream.

* Thunderstorms, and is *usually* found in and around mature thunderstorms.

* The jet stream, and is *infrequently* found within a layer between the height of the jet core and 5000 feet below the core, and approximately 50 to 150 miles towards the cyclonic (cold) side of the jet core.

* Cumuliform clouds, and is *infrequently* found in towering cumulus.

EXTREME TURBULENCE

Extreme turbulence is defined in relation to:

* The mountain wave when the strongest winds at mountain top level perpendicular to the ridge line are 50 knots or more. Extreme turbulence is *usually* found at low levels, lee-

ward of the mountains in or near the rotor cloud, if present.

* The mountain wave when the strongest winds at mountain top level perpendicular to the ridge line are 20 to 50 knots. Extreme turbulence is *infrequently* found at low levels, leeward of mountains.

* Thunderstorms, and is *frequently* found within a growing cell (indicated by hail, heavy rain, strong radar echo gradients or almost continuous lightning).

* Strongest forms of convection, wind shear or standing wave action, and is *usual*.

THERMAL CAUSES OF TURBULENCE

Vertical air movements or convective currents develop in air which is heated by contact with a warm surface. This heating from below occurs when cold air is advected (moved horizontally) over a warmer surface or the ground is strongly heated by solar radiation.

The strength of convective currents depends in part on the extent to which the earth's surface below has been heated and this, in turn, depends on the nature of the surface. Notice in Figure 152 that barren surfaces, such as sandy or rocky wasteland and plowed fields, are heated more rapidly than surfaces covered with grass or other vegetation. Thus, barren surfaces generally cause stronger convection currents. In comparison, water surfaces are heated more slowly. This difference in surface heating between land and water masses is responsible for the turbulence experienced when crossing shorelines on hot summer days.

When air is very dry, convective currents

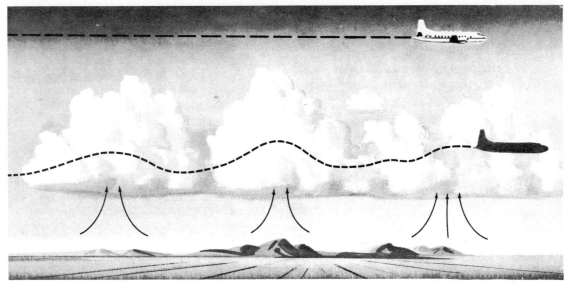

Fig. 153 Avoiding convective turbulence by flying above cumulus clouds.

may be present although convective-type clouds (cumulus) are absent. Figure 153 shows how pilots can avoid convective (thermal) turbulence by flying above the levels reached by convective currents. The general upper limits of convective currents are often marked by the tops of cumulus clouds which form in them when the air is moist, or by haze lines.

If each rising gust of air could be seen on a hot day, it would look something like a large bubble of oil rising through water. When these rising bubbles reach a level where the air is stable and they cannot rise farther, they must spread out sidewise. These cause gusts which may buffet an airplane from one side to another. Since the pilot must watch the instruments carefully in turbulence, he may experience intense muscular and eye fatigue in his continuous effort to maintain the attitude of the airplane. This condition can usually be avoided by a change in altitude, since the stable air higher up will be smooth.

Convective currents with their varying surfaces often affect an aircraft's final approach, as shown in Figures 154 and 155.

If the atmosphere is already unstable in an area where convection currents have been initiated by thermal effects, the convective currents will be sustained, or even accelerated, by the atmosphere. Instability can also be achieved by the advection of cold air into an area which may result in forming a layer of instability at the surface or aloft. Convective currents may then be inaugurated by this unstable situation. Note in the sketch, Figure 156, that the replacing of warm air aloft by colder air may result in the formation of a layer in which the lapse rate is more unstable, causing turbulence.

MECHANICAL CAUSES OF TURBULENCE

When the air near the surface of the earth flows over obstructions, such as irregular terrain (bluffs, hills and mountains) and buildings, the normal horizontal wind flow is disturbed and transformed into a complicated pattern of eddies and other irregular air movements. Figure 157 shows how surface obstructions cause mechanical turbulence. Note how the buildings or other obstructions near an airfield can cause turbulence (Figure 158).

The strength and magnitude of mechanical turbulence depend upon the speed of the wind, the roughness of the terrain (or nature of the obstruction), and the stability of the air. Stability seems to be the most important factor in determining the strength and vertical extent of the mechanical turbulence.

Mechanical turbulence has only minor significance when a light wind blows over irregular terrain. In such case, the turbulence is usually only a few hundred feet thick. When the wind blows faster and the obstructions are larger, the turbulence increases and extends to higher levels.

When strong winds blow approximately perpendicular to a mountain range the resulting turbulence may be quite severe. Associated areas of steady updraft and downdraft may extend to heights from 2 to 20 times the height of the mountain peaks. Under these

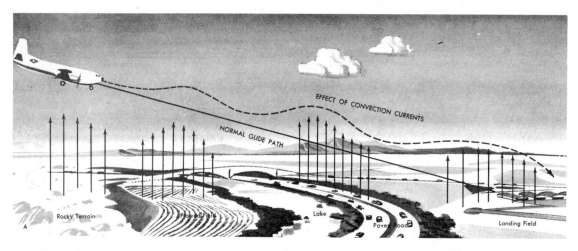

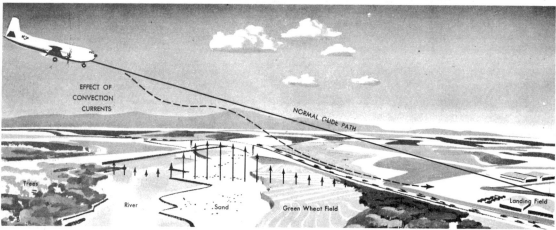

Fig. 154 Effect of convective currents on final approach. Vertical air currents may cause pilot to overshoot or under shoot—see Fig. 155—depending on strength and distribution of convection.

conditions when the air is stable, large waves tend to form on the lee side of the mountains and extend up to the lower stratosphere for a distance up to 100 miles or more downwind. These are referred to as standing waves or mountain waves and may or may not be accompanied by turbulence. Pilots, espe-

Fig. 156 Turbulence aloft caused by advection of cold air.

cially glider pilots, have reported that the flow in these waves is often remarkably smooth. Others have reported severe turbulence.

Note in Figure 160 that the mountains are at the right and the wind flow is from right to left. The cap cloud is shown on the mountain range crest to the right in the illustration. The roll cloud appears in the lower left-center portion of the sketch with the pile of lenticular (lens-shaped) clouds one above the other fanning out above and sloping a little windward toward the mountain range. Turbulence is most likely in the lee area up to the height of the mountains and again near the tropopause.

The airflow is fairly smooth and has a lifting component as it moves up the windward side of the mountain range. The windspeed gradually increases, reaching a maximum near the peak of the mountain. On passing the peak, the flow breaks down into a much more complicated pattern with downdrafts

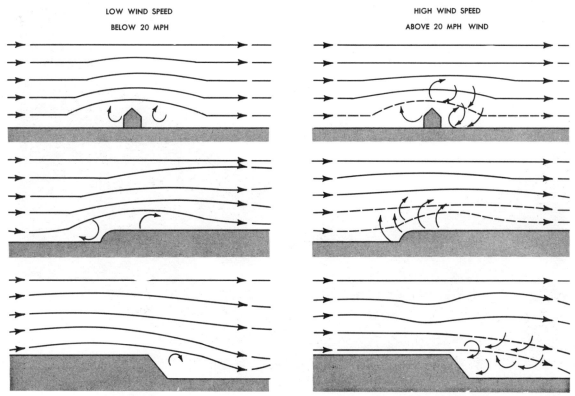

Fig. 157 *Surface obstructions cause eddies and other irregular wind movements.*

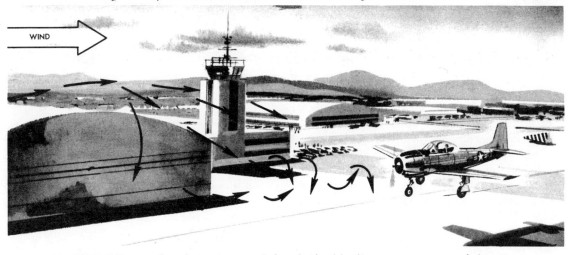

WIND

Fig. 158 *Building or other obstructions on windward side of landing area may cause turbulent air.*

predominating. Downwind, perhaps 5 to 10 miles from the peak, the air flow begins to ascend as part of a definite wave pattern which has been induced into the general flow by the mountain range. Additional waves, generally less intense than the primary wave, may form farther downwind.

The pilot is concerned, for the most part, with the first wave because of its more intense action and proximity to the high mountain terrain. Severe turbulence frequently extends to 150 miles downwind when the winds

are greater than 50 knots at mountain top level. Moderate turbulence often can be experienced out to 300 miles under the previously stated conditions. When the winds are less than 50 knots at mountain peak level, a lesser degree of turbulence will be encountered. Wave formation with roll clouds seems to require a certain degree of stability and a sufficient increase of wind speed with height in the middle troposphere.

Characteristic cloud forms, peculiar to wave action, provide the best means of visual iden-

Fig. 159 Typical wave clouds as seen from below (wind blowing from the right).

tification. Although the lenticular clouds in Figure 160 are smooth in contour, they may be quite ragged when the airflow at that level is turbulent. These clouds may occur singularly or in layers at heights usually above 20,000 feet. The roll cloud forms at a lower level, generally about the height of the mountain ridge. The cap cloud usually obscures both sides of the mountain peak. The lenticular clouds, like the roll and cap clouds, are stationary in position. The cloud formations themselves are a useful guide to the location of turbulence.

Some of the most dangerous features of the mountain wave are the turbulence in and below the roll cloud, the downdrafts just to the lee side of the mountain peaks and to the lee side of the roll clouds. The cap cloud must always be avoided in flight because of turbulence and concealed mountain peaks.

While clouds are generally present to forewarn the presence of mountain wave activity, it is possible for wave action to take place when the air is too dry to form clouds. This,

of course, increases the likelihood of flying into a wave unexpectedly.

TIPS ON FLYING THE MOUNTAIN WAVE

The six rules listed have been suggested for flight over mountain ranges where waves exist:

1. If possible, fly around the area when wave conditions exist. If this is not feasible, fly at a level which is at least 50 percent higher than the height of the mountain range. Be aware of the minimum safe altitude (terrain elevation plus 50%) during climb-out to cruising altitude and descents for landings.

2. Avoid the roll clouds since they are the areas with the most intense turbulence of the mountain wave.

3. Avoid the strong downdrafts on the lee side of mountains.

4. Avoid high lenticular clouds, particularly if their edges are ragged.

5. Do not place too much confidence in pressure altimeter readings near mountain peaks. They may indicate altitudes which are

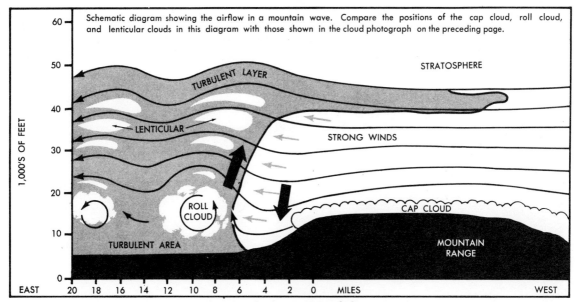

Fig. 160 Mountain wave turbulence.

more than 1,000 feet higher than the actual altitude.

6. Penetrate turbulent areas at air speeds recommended for your equipment.

FRONTAL CAUSES

Frontal turbulence is caused by the lifting of warm air by a frontal surface leading to instability and/or the mixing or shear between the warm and cold air masses. The vertical currents in the warm air are strongest when the warm air is moist and unstable. The most severe cases of frontal turbulence are generally associated with fast moving cold fronts. In these cases mixing between the two air masses as well as the differences in wind speed and/or direction (wind shear) add to the intensity of the turbulence.

Excluding the turbulence that would be encountered in any thunderstorms along the front, the accompanying diagram (Figure 161) illustrates the wind shift that contributes to the formation of turbulence across a typical cold front. As a general rule, the wind speed is greater in the colder air mass.

WIND SHEAR CAUSES

A relatively steep gradient in wind velocity along a given line or direction (either vertical or horizontal) produces churning motions (eddies) which result in turbulence. The greater the change of wind speed and/or direction in the given direction, the more severe the turbulence. Turbulent flight conditions are frequently encountered in the vicinity of the jet stream where large shears in the horizontal and vertical are often found. Since this

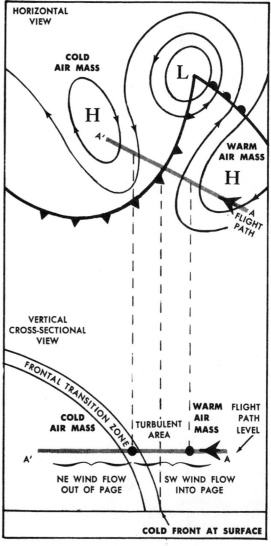

Fig. 161 Turbulence across a typical cold front.

Fig. 162 Turbulent air at the boundary between calm, cold air and moving warm air above.

type of turbulence may occur in perfectly clear air without any visual warning in the form of clouds, it is often referred to as clear-air turbulence (CAT). Clear-air turbulence is not necessarily limited to the vicinity of the jet stream and may occur in isolated regions of the atmosphere. For example: the turbulence in a mountain wave can also be classified as clear-air turbulence because the identifying clouds in the wave do not necessarily have to occur for the turbulence to be present.

A narrow zone of wind shear with its accompanying turbulence will sometimes be encoutered by pilots as they climb or descend through a temperature inversion. These inversions can and do occur anywhere from just above the surface to the tropopause. However, the tropopause is often the most significant inversion.

An extreme form of wind shear that is of considerable importance to aircraft landing and take-off operations is that which is associated with strong inversions near the ground. In the example, a pocket of calm, cold air has formed in the valley as a result of nighttime cooling, but the warmer air moving over it has not been affected appreciably. Due to the difference in airspeed between the two bodies of air, a narrow layer of very turbulent air is formed. An aircraft climbing or descending through this zone will encounter considerable turbulence (as well as changes in lift). Refer to Figure 162.

Moderate turbulence may be encountered, momentarily, when passing through the wake of another aircraft. On landing and departing the wake of aircraft produces turbulence in the approach path to and along runways. The turbulence in the wake of heavy aircraft is usually of concern to pilots of lighter aircraft. (Figure 163)

Fig. 163 Landing aircraft leave wake of turbulent air.

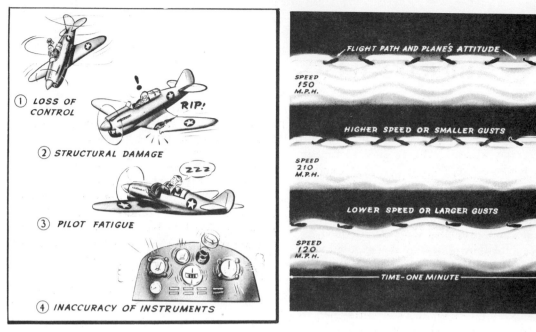

Fig. 164 Double pic of turbulent air flying.

HOW TO FLY TURBULENT AIR

In turbulent air the possibility of loss of control can never be discounted, even when an experienced pilot is doing the flying. The technique to be used in maintaining control will depend upon the cause of the turbulence, the strength and size of the individual gusts, and the type of airplane. Although there are no set rules for all conditions, the following procedures apply in a majority of cases:

1. Reduce air speed to maneuvering speed.

2. Guide the control lightly, maintaining direction and air speed within reasonable limits. Do not try to correct for each gust as it hits the plane. If the plane is properly trimmed, one gust will tend to counteract another and smooth out momentary variations. However, where successive gusts in the same direction cause a material change in attitude make a moderate correction.

3. Do not worry about your altitude too much if you are safely above obstructions. Fly attitude using the attitude indicator (artificial horizon).

4. Try to foresee where the air will be rough and at what altitudes. Avoid such regions by flying around them or by definite changes in altitude. It is particularly important to plan around "shear" lines where pressure areas meet.

5. Never exceed the maneuvering speed. The faster you go, the harder a gust will hit you. A strong gust encountered at high speed may result in structural failure.

The experienced pilot will avoid regions of moderate and severe turbulence whenever possible. He does not subject himself or his airplane to needless strains.

16 AIRCRAFT ICING

The less experience a pilot has had with icing conditions and the less he knows about ice accretion, the more likely he is to find it a problem. Aircraft icing is one of the major weather hazards to aviation. It affects an aircraft both externally and internally. In general, external icing (including what is called airframe and propeller icing) is defined as the accumulation of ice on the exposed surfaces of aircraft when flown through supercooled water drops (cloud or precipitation). Structural (external) ice reduces aerodynamic efficiency; in severe instances this efficiency is reduced to the point where an aircraft will no longer fly. Internal ice (intake, carburetor, or engine icing) literally reduces the breathing of the engine, which results in a loss of power.

The total effect of aircraft icing is:

* Loss of aerodynamic efficiency.
* Loss of engine power.
* Loss of proper operation of control surfaces, brakes and landing gear.
* Loss of outside vision.
* False flight instrument indications.
* Loss of radio communication.

Another important consideration is that icing limits the pilot's choice of maneuver. With reduction of maximum air speed, the ceiling of the aircraft is also rapidly reduced and the danger of loss of control is proportionately increased. The pilot may have no choice but to descend to a lower altitude. Over mountainous terrain or when the ceiling is low this is obviously hazardous.

In wind tunnel experiments it has been found that an ice deposit of 1/2-inch on the leading edge of some airfoils reduces the lifting power of the airfoil as much as 50% and increases the drag by an equal amount. Ice can form on an aircraft very rapidly, and there are recorded cases where two to three inches of ice have accumulated in a matter of minutes.

These facts indicate the importance of knowing something about aircraft icing. This chapter is intended to familiarize the pilot with the types of icing, weather conditions favorable to the formation of ice, hazards of aircraft icing, methods of preventing or reducing icing, and how to combat it after it has formed on or in the aircraft.

STRUCTURAL ICING
CONDITIONS CONTRIBUTING TO STRUCTURAL ICING

Two atmospheric conditions necessary for the formation of structural icing are: (1) presence of visible liquid moisture; and, (2) free-air temperature at or below freezing. Also, the aircraft-surface temperature must be colder than 0°C (32°F).

The temperature range in which ice and water coexist in clouds is generally from 0°C to minus 20°C. In the atmosphere, supercooled liquid water droplets are normally not in contact with ice and they continue to exist in a supercooled state at temperatures well below the freezing point of water. In fact, supercooled water droplets have been found at temperatures colder than minus 40°C and at altitude above 40,000 feet. Supercooled water droplets in the atmosphere are in a relatively unnatural state, even though they are frequently encountered. Any slight jar or contact with a surface will cause immediate freezing.

Air flow around aircraft surfaces will decrease the free air temperature in some cases as much as 2°C. This is of little consequence to the over-all icing problem but it explains why there are times when icing will take place when the outside free-air temperature gage indicates temperatures are slightly above freezing.

TYPES OF STRUCTURAL ICING

Aircraft structural icing consists of three basic types: clear (glaze), rime, and frost. Also, mixtures of clear and rime are common. The type of icing — clear or rime — is dependent primarily upon the water droplet size.

Clear ice is the most serious of the various forms of ice because it adheres tenaciously to the aircraft. It is formed by the relatively slow freezing of large, supercooled liquid-water droplets which have a tendency to spread out and assume the shape of the surface on which they freeze. As a result of the spreading of this supercooled water and its slow freezing, very few air bubbles are trapped within the ice, which accounts for its

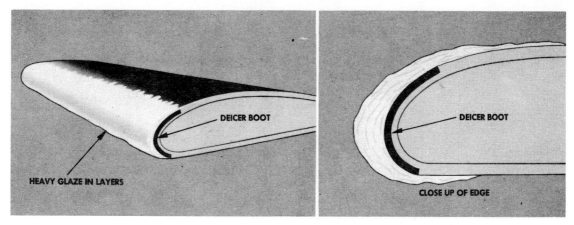

Fig. 165 Clear ice on an airfoil.

clearness. Although clear ice is expected mostly with temperatures between 0° and —10°C, it does occur with temperatures as cold as —25°C.

The atmospheric conditions which produce clear ice are encountered most frequently in cumuliform clouds. Clear ice also forms rapidly on aircraft flying in zones of freezing rain or drizzle.

Rime ice is formed by the instantaneous freezing of small supercooled water droplets upon contact with aircraft surfaces. Fast freezing can take place when the temperature is anywhere from 0°C to —40°C, but probably is most likely between 10° and —20°C. Since the individual droplets freeze in individual spheres on the airfoil and the freezing is instantaneous, a large amount of air is trapped within the ice. This gives the ice an opaque appearance, makes it very brittle and relatively easy to break off. Rime ice does not normally spread over an aircraft surface, but protrudes forward into the airstream along the leading edges.

Rime icing is most frequently encountered in stratiform clouds. In comparison to clear ice, rime ice is relatively easy to get rid of by conventional methods, even though it distorts the airfoil to a larger degree than does clear ice.

Frost is a thin layer of crystalline ice that forms on the exposed surfaces of parked aircraft when the temperature of the exposed surfaces is below freezing (while the free-air temperature may even be above freezing). This deposit of ice forms during night radiational cooling in a manner similar to the formation of hoar frost on the ground.

Frost is a most deceptive form of icing. It affects the lift-drag ratio of an aircraft and is a definite hazard during take-off. There is no such thing as a little frost on aircraft surfaces. The only definitions that apply are none or some, and some is too much.

Frost also forms in flight when a cold aircraft descends from a zone of freezing temperatures to a zone of above freezing temperatures and high relative humidity. The air is chilled suddenly to below freezing temperature by contact with the cold surfaces of the aircraft and sublimation (ice crystal formation directly from vapor state) occurs. Frost can cover the windshield or canopy and completely restrict outside vision. During penetra-

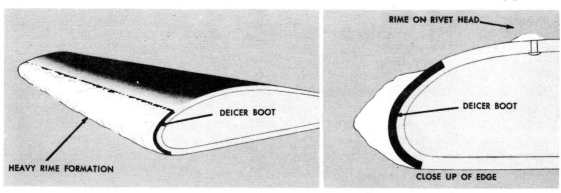

Fig. 166 Rime ice on an airfoil.

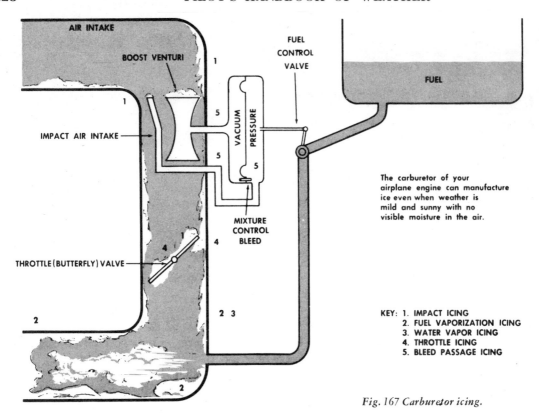

KEY: 1. IMPACT ICING
2. FUEL VAPORIZATION ICING
3. WATER VAPOR ICING
4. THROTTLE ICING
5. BLEED PASSAGE ICING

The carburetor of your airplane engine can manufacture ice even when weather is mild and sunny with no visible moisture in the air.

Fig. 167 Carburetor icing.

tion for landing, when visible moisture or high relative humidity exists, frost may form on the inner side of the canopy. This can be a definite hazard if no preventive action is taken.

Remove all frost from aircraft before take-off. Never be satisfied with less than an aerodynamically clear aircraft.

ICING HAZARDS

An important part in understanding the icing hazard is a knowledge of what is happening to the aircraft when an unexpected icing condition is encountered. When ice forms on an aircraft, it can affect its flying characteristics in several ways, depending on the part of the aircraft involved.

PROPELLER ICE

Ice deposited on the propeller of an aircraft may upset its aerodynamic efficiency, produce dynamic unbalance, and damage the airfoil structure when thrown off by centrifugal force. The pieces of ice thrown off are unequal in size and produce a great amount of dynamic unbalance. For example, the dynamic unbalance produced when one pound of ice is thrown from a propeller blade at a distance of four feet from the hub, when engine speed is 1,850 rpm, is 4,700 pounds. This is almost enough force, as one pilot expressed it, "to tear the engine out of the plane."

TAIL SURFACE ICE

A deposit of ice in any of its various forms on the tail curfaces of an aircraft is dangerous in that it may affect the longitudinal and directional control.

CARBURETOR ICING

Carburetor icing can occur under a wide range of temperatures and can result in complete engine failure. However, it is generally encountered only between 13° and 18°C outside air temperature in high relative humidity, rain, or overcast. Below 13°C, the danger of ice formation in the air is slight. However, once the ice begins to form in the carburator, the power can be seriously affected in less than a minute's time.

Ice may be formed in a reciprocating-engine induction system by two processes: (See Figure 167)

Impact ice is formed when supercooled water strikes a surface that is at sub-freezing temperature. It collects on scoop inlets, duct walls, carburetor-inlet screens, exposed metering elements, and other protuberances in the induction system. These ice formations may reduce the airflow and thereby reduce the engine power.

Throttle ice is formed because of the cooling effect of the evaporating fuel after it is introduced into the airstream. This is the carburetor ice that probably occurs most frequently in actual operation because it may form at carburetor air temperatures considerably above 0°C.

To combat carburetor icing, two basic types of carburetor heating systems are used in aircraft with reciprocating engines — the alternate air system and the preheat system. When flying in conditions conducive to induction system icing, use carburetor heat.

PITOT TUBE ICE

Ice in the pitot tube reduces its size and changes the flow of air in and around it. When this happens, the airspeed instrument becomes useless or unreliable. The pitot tube is heated electrically to remove the danger of icing.

SNOW

Snow sometimes clogs the oil radiators and the inside of the engine cowl, producing excessively high temperatures. Mechanical devices are available which reduce this effect to a minimum.

WINDSHIELD ICE

Ice becomes an annoyance when deposited on the windshield because it reduces visibility. It is particularly dangerous at time of landing. Ice forms simultaneously with windshield ice on the lens in front of the landing lights adding to the difficulty of safe landing at night. Ice may cover the windshield to the extent shown in the illustration below (Figure 168).

ICING HAZARDS ON THE GROUND

Frozen precipitation or frost accumulated on parked aircraft are hazards which greatly affect safe operation. If not adequately removed, they can cause considerable danger to the flight. It is common practice to place aircraft in a hangar until the ice melts, or to completely clean wing and tail surfaces prior to flight. If hangared for deicing, keep in mind that when the aircraft comes out of the hangar there will probably still be some water left on the surfaces and it will ice again if freezing temperatures exist.

Another serious condition arises from the presence of water or mud on ramps and runways during freezing temperatures. Water blown or splashed on the aircraft forms an ice cover on the under side of flaps and control surfaces. Mud and water may freeze in the brakes and in the landing gear mechanism to the extent that proper operation is impaired. Be alert and conduct your ground operation accordingly.

Ice and snow on runways are among the weather elements that affect braking action of aircraft. Brake-friction ratio varies widely with different aircraft, and each pilot must be aware of his own aircraft's characteristics.

DEICING AND ANTI-ICING

Some aircraft are equipped with deicing and/or anti-icing equipment. There are three common methods for eliminating ice: (1) mechanical, (2) fluids, and (3) heat.

Fig. 168 Windshield ice.

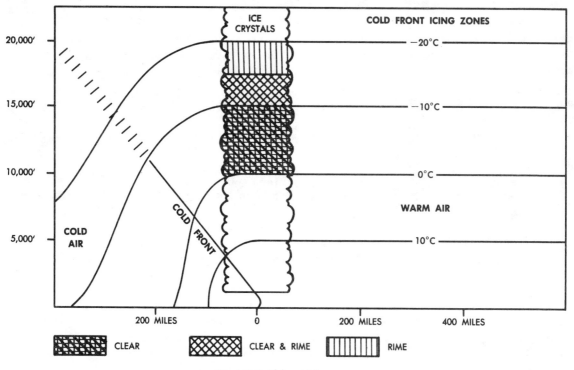

Fig. 169 Cold front icing zones.

Mechanical — The leading edges of wing and tail surfaces of some aircraft are equipped with rubber skins, or boots, that normally assume the contour of the airfoil. Compressed air is cycled through ducts in these rubber boots causing them to swell and change shape. The stress produced by the pulsating boots cracks the ice and the air stream peels it off.

Anti-icing fluids are used on rotating surfaces where centrifugal force spreads the fluid evenly. Such fluids are effective agents because the fluid helps prevent ice from adhering to the surfaces and the centrifugal

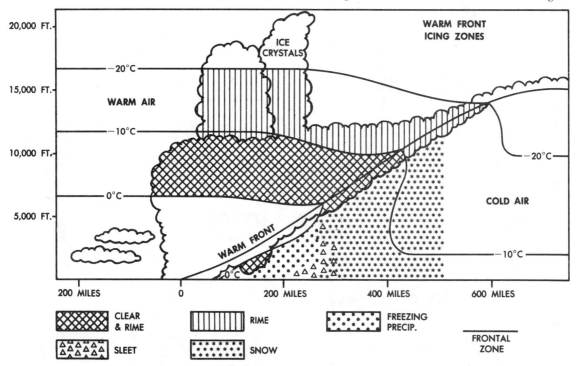

Fig. 170 Warm front icing zones.

force throws off the ice as it forms.

Heat application capability to wing, props, and tail surfaces is installed in many newer aircraft. The icing areas are heated by electrical means or by hot air which is piped from the engine manifold. This process has frequently given rise to the name hot wing.

WEATHER CONDUCIVE TO AIRCRAFT ICING

The pilot should anticipate and plan for some type of icing on every flight conducted in clouds with temperatures colder than freezing. He should be familiar with the icing generally associated with different atmospheric conditions.

AIR MASS ICING

Stable air masses often produce stratus-type clouds with extensive areas of relatively continuous icing conditions. These clouds represent a potentially hazardous icing condition if the pilot is prevented or delayed from making the necessary changes in flight altitude. Studies indicate that an individual cloud layer seldom exceeds 3,000 feet in thickness. However, situations are encountered where multiple layers of clouds are so close together that visual navigation between layers will not be feasible because of the variations in the tops and bases of the clouds. In such cases the maximum depth of continuous icing conditions will rarely exceed 6,000 feet. Generally, the pilot should expect rime icing in stratus-type clouds when the free-air temperature is below 0°C.

Unstable air masses generally produce cumulus clouds with limited horizontal extent of icing conditions, but the pilot can expect the icing to become more severe at higher altitudes in the clouds. The concentration of supercooled water droplets in cumulus clouds may be several times greater than that existing in stratus clouds, thus resulting in much greater icing, but of shorter duration. Clear ice is generally associated with cumulus-type clouds.

Pilots can generally expect more icing while flying over mountainous terrain under icing conditions than over other types of terrain with the same atmospheric conditions. When flight planning, anticipate heaviest icing around 5,000 feet above the mountain tops, and/or in the temperature range at and just below freezing —0°C to —10°C, or vicinity.

FRONTAL ICING

Cold fronts and squall lines generally have a narrow weather and icing band and the majority of the time the associated clouds will be the cumuliform type. The depth of the icing zone will often be about 10,000 feet and the type will be predominantly clear. Icing is very severe when the overrunning or lifted warm air is unstable. In Figure 169 the icing zone is about 100 miles wide and covers altitudes of about 10,000 feet.

Different types of icing encountered in the clouds will generally be: clear icing, where the temperature ranges from 0° to —10°C; a mixture of clear and rime, where the temperature ranges from —10°C to —15°C; rime icing, where the temperature ranges from —10°C to —20°C, or less. However, icing of all types may occur at much colder temperatures; —25°C for clear ice, and —40°C for rime ice are accepted lower limits.

Warm fronts and stationary fronts generally have a wide weather and icing band with stratiform clouds. The vertical depth of the icing zone will generally be about 10,000 feet, and the type of icing will be predominantly rime. If the overrunning warm air is unstable, there will also be cumuliform clouds creating a more severe icing hazard. Note in Figure 170 that the icing zone is 500 miles wide and covers about 17,000 feet in altitude. Freezing precipitation may be extensive.

The very critical freezing precipitation area is where water is falling from warm air above when the flight level temperature is below freezing. In this case clear ice would be encountered and the evasive action is to climb to an altitude where the temperature is above freezing. Icing in the clouds may follow this basic pattern — a mixture of clear and rime ice from 0°C to —10°C; rime from —10°C to —20°C. Again rime icing may be expected occasionally at temperatures below —20°C.

Occluded fronts often produce a wide weather and icing band with both stratiform- and cumuliform-type clouds. The depth of the icing zone will often be about 20,000 feet, or double the depth of icing zones with other type fronts. The types of icing will be clear, mixed and rime and if the air masses involved are unstable, aircraft icing may be very severe.

In Figure 171 notice that the freezing precipitation area is quite extensive, so keep in mind that when the water comes from warmer air above, the icing will be most hazardous.

The types of icing associated with the temperature ranges will generally be: clear, 0°C

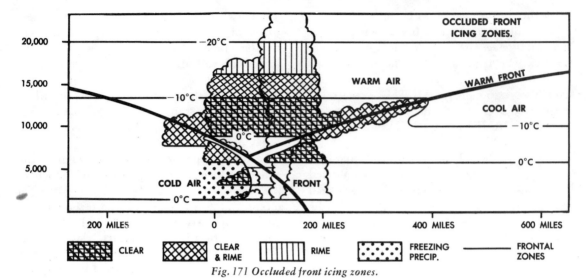

Fig. 171 Occluded front icing zones.

to —10°C; a mixture of clear and rime —10°C to —15°C; rime, —15°C to —20°C, with possible rime icing at lower temperatures. Notice the warm air pocket between the altitudes of 6,000 and 9,000 feet. Know the icing zones, and plan your flight accordingly.

THUNDERSTORM ICING

The icing areas and the type of icing in the different stages of a thunderstorm are shown in Figure 172.

In the cumulus cloud that is developing into a thunderstorm, the cloud particles above the freezing level will, to a large extent, be liquid. The icing will be severe due to the large number of liquid water droplets. When the top of the cloud begins to reach —20°C, ice crystals begin to form and aircraft icing will be reduced, as shown.

In the updrafts in the mature cell, the cloud particles are mostly water where the temper-

ature is warmer than —10°C. Water droplets and ice crystals are mixed between —10° and —20°C; and ice crystals will predominate at temperatures colder than —20°C. In the downdrafts, the cloud will be mixed ice and water from 0°C to —10°C, and mostly ice crystals when the temperature is colder than —10°C.

In the dissipating cell ice crystals predominate (Section C). However, in a shallow layer near the freezing level there is a mixture of water droplets and ice crystals where aircraft icing may take place.

It is very difficult to tell the stage of a thunderstorm by visual observation, so the pilot should anticipate icing at all altitudes where the temperature is at or below freezing. Remember that the worst icing conditions are encountered at and just above the freezing level. Icing can occur also in the cirrus anvils

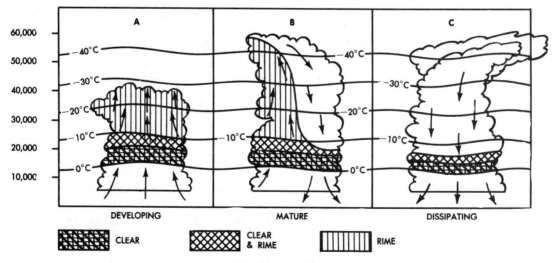

Fig. 172 Thunderstorm icing zones.

streaming off the upper parts of thunderstorms because some water drops are carried out into the anvil and do not always freeze immediately.

WEATHER REPORTS

Weather personnel can not generally observe icing. They rely on pilot reports. Forecasters can forecast the probable maximum intensity of icing that may be encountered during a flight but not necessarily the intensity of icing that will be encountered by a particular aircraft. Many variables bear upon this problem. It is the pilot's responsibility to obtain a complete weather briefing to minimize any icing hazard and all other hazards to his flight.

17 SHOWERS AND SQUALLS

Showers and squalls, because they are relatively small in area and usually surrounded by clear air, are too often regarded lightly by inexperienced pilots. This is a dangerous attitude, and the history of aviation records its disastrous results. Some of the most severe turbulence and worst icing conditions are encountered in small, localized squalls.

Where rain or snow is falling, ice accretion will almost always be found at some level. Where heavy rain is occurring, and particularly in showers, turbulence will also be found at the level at which the precipitation is originating. This is because the physical process which forms rain requires the presence of ice crystals and water droplets at subfreezing temperatures in the same region of a cloud. These conditions also favor ice accretion. To form heavy rain extensive lifting of the air is necessary, implying the presence of strong vertical currents and turbulence. However, it would be a mistake to assume that icing and turbulence are not present simply because precipitation is not occurring. Both icing and turbulence frequently occur without rain or snow or any apparent threat of them.

WHYS OF SHOWERS AND SQUALLS

The distinction usually made between a shower and a squall is a fine one based upon the absence or presence of strong, gusty surface winds and the intensity of precipitation. These are characteristics which can best be observed from the ground, but they are likely to be meaningless for the pilot. Showers and squalls will, therefore, be discussed together in this section.

Both showers and squalls occur in conditionally unstable air, that is, in air which is stable when unsaturated but becomes unstable when saturation occurs. Either type of storm may be produced by surface heating or released by such localized upward displacement of air as occurs over a mountain ridge. Except for a difference in intensity, these two types of storms present similar flight conditions and require the same flying technique.

The intensity of a local storm will be governed by the amount of moisture available and by the amount of energy released when this moisture condenses. The amount of energy released will be large when a steep lapse rate, through a comparatively deep layer of air, indicates that air can rise unstably for a considerable distance. The production of cloud and precipitation will proceed rapidly under such conditions, and the resulting weather is described as a squall. A layer of unstable air at least 10,000 feet thick is usually necessary before a heavy squall can be produced.

If upper air observations of temperature and humidity are available, the forecaster can usually predict the occurrence and intensity of squalls or showers by noting the steepness of the lapse rate and the amount of water vapor available to provide energy for the storm.

IT'S A QUESTION OF TEMPERATURE

Unstable rising air currents may be started through heating of the surface air by contact with sun-warmed ground or by upward displacement of a portion of the air. When both factors are active, as on a hot day over the barren mountainous terrain from central Texas westward, violent squalls may be produced. Such storms often remain stationary over a given hill or peak throughout the period of their activity.

The slope of most frontal surfaces is too slight to provide the upward thrust necessary to initiate the formation of squalls; but cold fronts, having relatively steep slope and high speed, form squalls more readily then warm fronts. Showers are more common along warm fronts.

One physical difference between a squall and a shower is in the vertical distance between the base of the cloud and the level at which ice crystals begin to form rapidly (—10°C). Most precipitation is produced at or above this level, hence a heavy shower will develop in and fall from a smaller cloud when the ice-crystal level is low. If the cloud cannot go above this level, no precipitation, or only a light shower of large raindrops, may result.

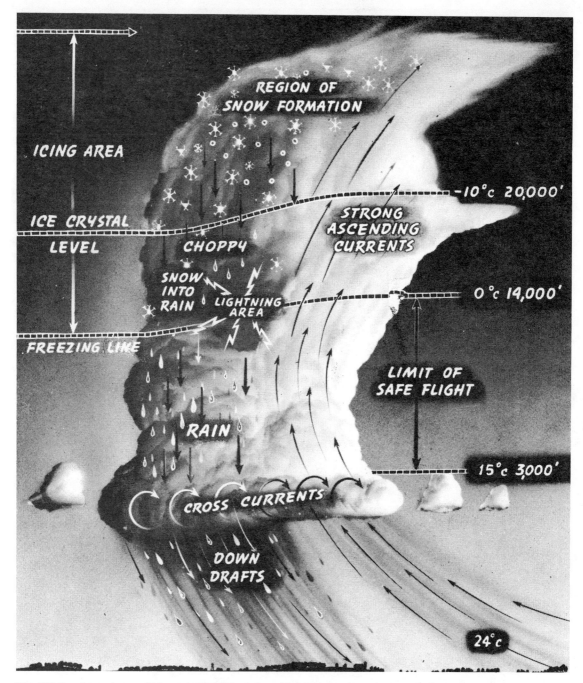

Fig. 173 Structure of a squall in summer: The rain squall should be circumnavigated if possible. If necessary, flight should be conducted through the cloud mass in the rain area or above the ice-crystal level. The intermediate level should be avoided to eliminate the hazard of ice and lightning.

DANGERS TO EXPECT AND AVOID

Ice conditions will prevail in the upper regions of all squalls and most showers. The temperature gage or upper air temperature observations will tell the pilot the altitude of the freezing level, below which icing will not occur.

Since the intensity of precipitation from a shower or squall is an indication of how much moist air is being lifted, it is also a warning of how much turbulence may be expected within the storm. The more unstable the air, the heavier will be the precipitation and the more intense the turbulence. The more rapidly the air rises, the greater will be the quantity of liquid water carried above the freezing level, and the more severe will be the resulting icing condition. Therefore, ice accretion is more severe in squalls than in showers.

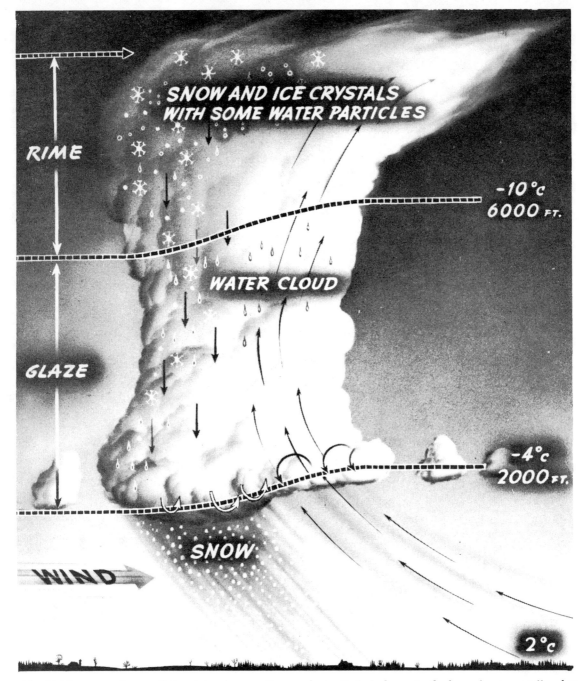

Fig. 174 Structure of a squall in winter - ice accretion can be particularly heavy in the base of snow squalls when the temperature is between 0° and —10°C. Winter squalls may be flown through at upper levels and only a light to moderate amount of ice accumulated. Turbulence is usually of the same degree.

KNOW THEIR LIFE HISTORIES

A squall or shower has a definite life history. It grows, reaches maturity, becomes old and weak, and finally dies. The intensity of turbulence and icing conditions is directly related to this cycle. The age factor must always be considered. A pilot who has formed his ideas of the intensity of hazards in squalls from experience with old or decadent storms may get a dangerous or even disastrous sur-

prise when he meets a squall at the peak of its maturity.

Squalls and showers reach their peak of violence early in life, just when the cloud attains its maximum height and rain or snow begins to fall from the base. Thereafter, the intensity of turbulence and ice accretion slowly decreases. This decay is usually indicated by a fading of the sharp outlines of the cumuliformed summit of the storm cloud. This fading usually progresses downward from the upper portion of the cloud.

The energy that sustains a squall or shower is derived solely from the air fed into it at its base. No appreciable quantity of air enters the cloud at intermediate levels. Consequently, such storms have a tendency to develop rapidly and to be short-lived, soon declining when the supply of warm air at the ground is exhausted or cut off. Showers and squalls, therefore, occur as localized storms.

The heavy precipitation which occurs as air sinks within the storm, interrupting the the squall or shower reaches maturity carries colder temperatures downward. Locally chilled progress of ascending air and diminishing the intensity of the storm's development. When precipitation is encountered in a squall or shower, violently choppy and bumpy air, as well as drafts, will usually be found.

BETTER GO ANOTHER WAY

Near the base of the cloud the ascending air about to enter the storm is strongly repulsed by the cold descending air, and strong crosscurrents are produced. This region of a squall or shower should be avoided if possible.

When precipitation falls from the base of a cloud, the unsaturated air beneath is cooled by contact and evaporation. Being then colder than the surrounding air, it will fall unchecked to the ground, producing a strong downdraft.

Because the onset of heavy precipitation from a squall or shower is usually sudden and the area of activity small, these downdrafts are often strong enough to throw an airplane out of control or possibly carry it into the ground.

WHERE LIGHTNING STRIKES

Records reveal that the majority of lightning strikes involving aircraft have occurred in squalls or showers when the airplane was flying at a level where the temperature was near freezing. This region should always be avoided. Just below the freezing level the snow which formed in the upper portion of the cloud changes to sleet and rain. During this process a concentration of electrical charge occurs. It usually dissipates by natural means; but an airplane moving through the region may serve as a conductor to shorten the electrical circuit between oppositely charged portions of the cloud, causing a lightning discharge.

Because the precipitation usually begins suddenly in a squall or shower, the electrically charged region is formed rapidly and the charge is accumulated faster than it can dissipate by natural means. The lightning danger is, therefore, greatest when the storm is at its peak of activity.

THE LIGHTNING HAZARD AND PRECAUTIONS

There are usually several warning signs when a lightning discharge is about to occur and the pilot, being forewarned, can take proper precautions to avoid it or minimize its effect. When flying through squalls, showers, or even towering cumulus clouds, a lightning discharge should be anticipated whenever:

1. The temperature is between 5° and —10°C.
2. Mixed rain and snow is encountered.
3. Severe precipitation static occurs in the radio.
4. Corona forms on the propeller or other parts of the plane.

To avoid or minimize the effects of a discharge:

1. Seek a lower altitude.
2. Reduce speed to maneuvering speed.

3. Keep eyes focused on instruments, which should be brightly lighted at night.
4. Have automatic pilot adjusted for level flight and ready to engage. You may be temporarily blinded and the compass made inaccurate by a lightning discharge.

Fig. 175 *The major lightning hazard is temporary blindness by the brilliancy of a nearby discharge. Guard against this danger.*

MOUNTAINS AND FOOTHILLS ARE BREEDERS

The upward thrust of air required to initiate a squall or shower can be produced by an abrupt slope of the terrain, forcing air flowing over it upward to the condensation level. Squalls and showers are, therefore, common over mountains whenever conditionally unstable air moves across them.

A storm over a ridge or mountain will generally remain stationary instead of drifting with the air flow. It grows continuously on its upwind side, becoming old and dissipating progressively on its downwind side in the lee of the mountain or ridge. The most rapid cloud growth. with its turbulence and icing, will occur over the summit, and the heaviest precipitation on the lee side. When a downdraft occurs on the lee side of a hill in conjunction with a shower or squall, the strength of the downdraft is added to the normal down-slope flow over the obstruction. Downdrafts over rough terrain are, therefore, more intense than over level ground and should be avoided.

In the same way, rising currents initiated in mountainous regions by heating of the ground are intensified on the windward slopes of mountains so that updrafts and turbulence will be accentuated there.

Foothills and low mountain ranges whose crests are below the cloud bases are notorious for violent squalls. The Allegheny Mountains in the eastern United States are an excellent example.

Idealized diagrams can be made showing the structure and development of squalls and showers. These structures should be studied carefully and the hazardous regions of the storm memorized.

FLIGHT RULES TO BRING YOU THROUGH

From study of the structure and characteristics of squalls and showers, a number of flight rules can be derived. If a pilot learns them and applies them, the hazard of flight in these storms can be virtually eliminated. First let us review the problems to be met. They are:

1. TURBULENCE.
    Always present, most severe during the early part of the storm's activity. In the region of ascending currents, strong drafts predominate.
    In the region of falling snow or rain the air is choppy and bumpy.
    In the base of the cloud buffeting turbulence may be expected.
    Downdrafts are found in the region of precipitation below the base of the cloud and may occur elsewhere.
2. ICE.
    Moderate ice is to be expected in all portions of the cloud where the temperature is below freezing.
    Glaze (clear ice) usually occurs in regions of ascending air.
    Rime usually occurs in the region of precipitation.
3. LIGHTNING:
    Always possible near the freezing level.
4. PRECIPITATION.
    Rain or snow, depending on the temperature. Moderate to heavy, localized, and usually short of duration.

RECOMMENDED FLIGHT PROCEDURES

1. Fly around all squalls and showers at a level above the base of the cloud.
2. If it is absolutely necessary to go through the storm without de-icing equipment, fly at an altitude above the turbulent base but below the freezing level. Before entering the cloud:
    Apply carburetor heat:
    Secure safety belt.
    Reduce speed to the maneuvering speed.

    Establish heading on gyro compass.
    If your plane is equipped with de-icers, perform the above operations upon entering the cloud.
    Hold heading.
    Maintain lateral balance by easy control.
    Ride with the vertical currents.
3. Do not:
    Dive out of a squall or shower.
    Dive under a squall or shower to maintain ground contact.
    Attempt to turn out of a storm, except as a last resort.
    Fly at a level where the temperature is near freezing.

A thunderstorm is formed by the same processes as a squall and has many of the squall's characteristics. However, the thunderstorm has greater intensity, larger size, and more lightning and thunder.

Flight through a thunderstorm should be avoided whenever possible. Because there are times when there is no alternative, it is essential that a pilot know the various types of thunderstorms so that he may select a safe flight path. On his first flight through a thunderstorm, the average pilot will not find any path much to his liking. However, if he keeps cool, he will discover that his fear (and that is what it is) does not result from his actual predicament at the moment. He is afraid of imaginary dangers he visualizes ahead. Passage through the average storm, regardless of type, is rough. Rain will be heavy, lightning brilliant, and turbulence variable, usually moderate to severe. These conditions demand constant alertness, and the pilot must also be on the watch for the hazards of icing and hail. The dangers which he may encounter in a thunderstorm are loss of control, loss of power, and damage by hail, lightning, or extreme turbulence. These dangers can all be foreseen and precautions can be taken to minimize their effect.

If the day comes when you must fly through a thunderstorm, bear in mind that nothing will help you more than the exercise of your own good judgment. There has never been a thunderstorm that a good plane piloted by a good man couldn't lick. So, if the wind bangs you around, if lightning blinds you and hail threatens to bust up your windshield, keep your head, use your best judgment and fly your course. You'll reach your destination — or alternate. You'll get "there."

TYPES OF THUNDERSTORMS

In general thunderstorms have similar physical features, regardless of location or time. However, they do differ in intensity, degree of development, and in associated

Fig. 176 Fly it out.

Fig. 177 Warm front thunderstorm.

Fig. 178 Cold front thunderstorm.

weather such as hail, turbulence, electrical discharges, and high or gusty winds. Thunderstorms are generally classified according to the manner in which the initial lifting action is accomplished.

The different types according to this criterion are:
* Frontal Thunderstorms
* Warm Front
* Cold Front
* Prefrontal (Squall Line)

* Occluded Front
* Air Mass Thunderstorms
* Convective
* Orographic
* Nocturnal

FRONTAL THUNDERSTORMS
WARM FRONT THUNDERSTORMS

Warm front thunderstorms usually occur when the warm, moist air overrunning the cold mass of retreating air is unstable. Notice the overrunning effect in Figure 177. Because

Fig. 179 The line squall - this type of storm can be exceptionally rough, but when flown through at right angles is usually of short duration. Do not parallel it flight just over the roll cloud or on top if conditions permit, is the proper procedure.

of the gentleness of the warm frontal slope, stratiform clouds often surround and obscure the thunderstorm clouds. For example, an aircraft flying from the right side of the illustration toward the left would fly from an area of relatively smooth conditions under the warm front into turbulence in a matter of seconds. Showery precipitation encountered in flight through warm fronts should be considered an indication of cumulus activity.

COLD FRONT THUNDERSTORMS

Cold front thunderstorms, as seen in Figure 178 occur in the warm air near the frontal surface. A continuous line parallel to and along the frontal surface is the distinguishing feature. Because most of the thunderstorms are visible, they are easy to recognize while approaching the front from any direction. The bases of cold front thunderstorms are usually lower than the warm front type. They are most active during the afternoon and are generally more violent than the warm front type.

PREFRONTAL OR SQUALL LINE THUNDERSTORMS

The prefrontal squall line is found 50 to 300 miles in advance of a cold front, and is generally parallel to it. The squall line is approximately 150 to 300 miles in length, although not necessarily continuous, and as much as 35 miles in width. This band of thunderstorms is very similar to a cold front line, but is often more violent and moves faster. The cloud bases are often lower and the tops higher than most thunderstorms. The most severe conditions, such as heavy hail showers, destructive winds, and tornadoes are generally associated with squall lines.

OCCLUDED FRONT THUNDERSTORMS

Thunderstorms occurring with cold front-type occlusions are similar to the cold front line of storms, but are generally not as exten-sive nor as severe. They are concentrated at the peak of the warm sector for a short distance along the newly formed occlusion.

On the other hand, occluded front thunderstorms are more often associated with the warm front-type of occlusion. In this case they occur along the upper cold front and are inaugurated by the rapid lifting of the warm air. They are similar to warm front type thunderstorms but often are more severe. As in the case of the warm front thunderstorm, the occluded front thunderstorms are often embedded in stratiform clouds and give little or no visible warning of their presence.

AIR MASS THUNDERSTORMS

Air mass thunderstorms have two basic characteristics: (1) They generally form within a warm, moist air mass, and (2) They are generally isolated or scattered over a large area.

CONVECTIVE THUNDERSTORMS

Convective thunderstorms occur with a greater frequence than any other type. They occur over land or water in most areas of the

Fig. 180 Convective thunderstorm.

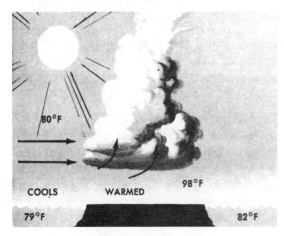

Fig. 181 Convective thunderstorm at day.

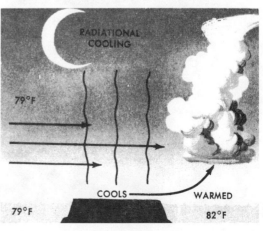

Fig. 182 Convection thunderstorm at night.

world, and are very common in the temperate zone during the summer months. The initial lifting action for these thunderstorms is provided by the convective currents produced by the heating of the lower layers of the air that are in contact with warm land or water masses.

Convective cumulus clouds normally form over land during the afternoon when the earth is receiving the maximum heating from the sun. When the air is very unstable and contains sufficient water vapor, the cumulus clouds may develop into thunderstorms. Convective thunderstorms also develop over coastal regions during the afternoon when cool, moist air from the water is heated as it moves over the warmer land surface.

Convective thunderstorms form over water at night as cool air from the land is advected over the warmer water surface. These thunderstorms are common in coastal areas, and are either isolated or scattered over a large region. They are easily recognized by their towering mass, lightning, and accompanying rain showers.

OROGRAPHIC THUNDERSTORMS

Orographic thunderstorms form when moist, unstable air is forced aloft by mountainous terrain. These storms develop rapidly and may cover large areas. They frequently remain stationary on the windward side of mountains or hills for many hours. They can form as single storms over peaks, or as a line of storms along a mountain range. Hail is common in these thunderstorms when they develop along high mountain slopes, such as the Rockies.

It is difficult to identify orographic thunderstorms when they are approached from the windward side of the mountains because they are often embedded in stratiform clouds. However, when they are approached from the lee side of the mountains, identification is usually easy.

It is not advisable to attempt a flight through the lower portion of an orographic thunderstorm because they almost always enshroud the tops of mountains and hills.

NOCTURAL THUNDERSTORMS OF THE MIDWEST

This is a peculiar type of air mass thunderstorm which occurs frequently late at night and early in the morning during the late spring and summer in the central plains area of the United States from the Mississippi Valley region westward. Its occurence is generally considered to be possible when a relatively

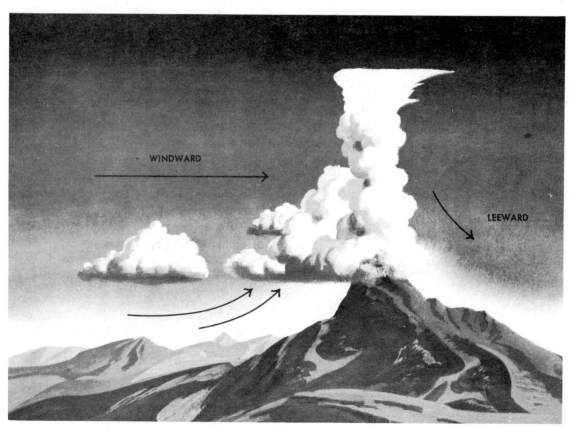

Fig. 183 Orographic thunderstorm.

moist layer of air exists aloft. The trigger mechanism that sets off this type is not well understood — it may result from advection of instability at middle levels formed farther west during the afternoon, or by cooling of cloud tops from nocturnal radiation, or complex diurnal variations in the wind structure. (Frontal and squall line thunderstorms can, of course, occur at any time — day or night.)

STRUCTURE OF THUNDERSTORMS

The fundamental structural element of the thunderstorm is the unit of convective circulation known as the convective cell. A thunderstorm usually contains several of these cells, each varying in diameter from about one to five miles, and each in a different stage of development. However, the circulation in each cell is generally independent of that in surrounding cells in the same storm.

Each thunderstorm progresses through a cycle which consists of three stages: (1) the cumulus stage, (2) the mature stage, and (3) the dissipating or anvil stage.

CUMULUS STAGE

In its initial stage of development, the thunderstorm is a cumulus cloud. As the development progresses, several cumulus clouds may unite to form a single thunderstorm. Only a small percentage of the cumulus clouds develop into mature thunderstorms. The main feature of the cumulus or building stage is the updraft, as you can see in Figure 183. This updraft prevails throughout the entire cell. During the cumulus stage, rain usually does not fall because the water droplets are carried upward or remain more or less suspended in the ascending air currents.

MATURE STAGE

When the water droplets in the cumulus cloud, as here shown, grow to the extent that they can no longer be supported by the updrafts they begin to fall. As the raindrops fall they drag air along with them. This drag is a major factor in the formation of downdrafts which characterize each convective cell in the mature stage. Downdrafts form in the middle regions of the cell and gradually increase, first vertically and then horizontally. When the downdraft air currents reach the surface, they spread out horizontally and produce strong and gusty surface winds. All of the hazards associated with thunderstorms seemingly reach maximum intensity during the mature stage.

Rain at the ground generally indicates the transition from the cumulus stage to the ma-

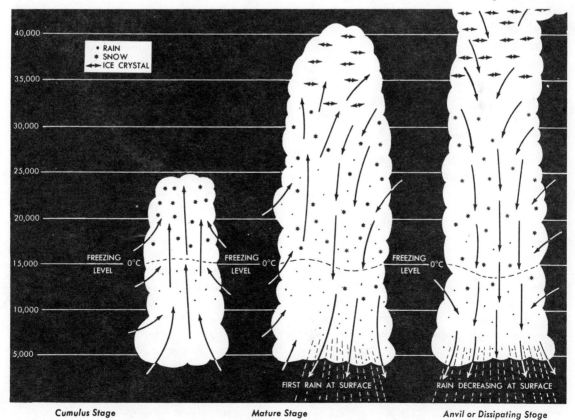

Cumulus Stage   Mature Stage   Anvil or Dissipating Stage

Fig. 184 Thunderstorm structure stages.

Fig. 185 A huge storm cloud looms 50,000 feet above western Oklahoma. It is estimated that the violence created by this storm was 50 times more than the bomb that struck Hiroshima.

ture stage. When the mature stage is reached, the cell usually has attained a height of more than 25,000 feet (60,000 feet is not unusual).

DISSIPATING OR ANVIL STAGE

Throughout the life span of the mature cell the downdrafts continue to develop, and spread vertically and horizontally. Also, as the downdrafts are developing, the updrafts are continually dissipating. As a result of this action, the middle and lower portions of the thunderstorms ultimately become an area of downdrafts.

Because of the heating and drying process produced by the downdrafts, the rainfall gradually ceases and the thunderstorm begins to dissipate. During this stage the lower level of the thunderstorm frequently becomes stratiform in appearance, and the top of the cell develops the characteristic anvil structure.

WEATHER WITHIN THE THUNDERSTORM

Thunderstorms are characterized by turbu-lence, moderate to extreme up- and down-drafts, hail, icing, lightning, precipitation and, under most severe conditions, (in certain areas) tornadoes.

TURBULENCE (DRAFTS AND GUSTS)

Downdrafts and updrafts are vertical currents of air which are continuous over many thousands of feet of altitude and are continuous over horizontal regions as large as a thunderstorm. Their speed is relatively constant as contrasted to gusts, which are smaller-scale discontinuities or variations in the wind flow pattern extending over short vertical and horizontal distances. Gusts are primarily responsible for the bumpiness (turbulence) usually encountered in cumuliform clouds. A draft may be considered as a river flowing at a fairly constant rate, whereas a gust is comparable to an eddy or other type of random motion of water in a river.

Drafts displace aircraft vertically. If the pilot does not fight or attempt to counteract such a vertical displacement too much, the

Fig. 186 Hail damage to aircraft.

possibility of structural damage or stalling is minimized.

Turbulence appears to increase in intensity with altitude in thunderstorms to within 5,000 or 10,000 feet below the tops of the clouds. Although turbulence is usually at a minimum near the bases and tops of the cloud, in some cases it may be severe in these areas.

ICING

Where the free-air temperatures are at or below freezing, icing should be expected in flights through thunderstorms. In general, icing will be associated with temperatures from 0°C to —20°C. Most severe icing occurs from 0°C to —10°C. In scattered thunderstorm areas icing will not present too serious a problem because the flight time in the storm will be relatively short. In areas of numerous thunderstorms the icing problem may be serious with prolonged exposure to icing conditions. Icing in the anvil tops is also sometimes encountered.

HAIL

Hail is primarily encountered during the mature stage, however, it is frequently carried aloft in the updrafts, cast out into the clear air away from the clouds, and appears to fall from the anvil or side of the cloud. Thus, occasionally, hail is encountered in the clear air near a thunderstorm (up to 5 miles out).

Although encounters by aircraft with large hail are not too common, hail larger than $\frac{1}{2}$- or $\frac{3}{4}$-inch in diameter can damage an aircraft in a very few seconds. Often the airframe is damaged significantly. The conclusion is reached that many if not most mid-latitude thunderstorms contain hail sometime during their life-cycle, with most hail occurring during the mature stage. In subtropical and tropical thunderstorms hail seldom reaches the ground, and it is generally believed that they contain less hail aloft than middle and high latitude storms. However, this does not mean that there is no danger from hail in the tropics.

Forecasting possible hail has improved tremendously, but with every thunderstorm possessing unknown individual qualities, the pilot should be prepared for hail in any thunderstorm.

RAIN

In thunderstorms a pilot can expect to encounter considerable quantities of visible moisture which may not necessarily be falling to the ground as rain. These water droplets may be suspended in, or moving with, the updrafts. Rain will be encountered below the freezing level in almost all penetrations of fully developed thunderstorms. Above the freezing level, however, there is a sharp decline in the frequency of rain.

There seems to be a definite correlation between turbulence and precipitation. The intensity of turbulence, in most cases, varies directly with the intensity of precipitation. This relationship indicates that most rain and snow in thunderstorms is held aloft by drafts.

LIGHTNING

During thunderstorm penetrations, lightning has proved to be a minor hazard. Aircraft struck by lightning generally receive small punctures in the skin. However, experience has shown that aircraft with large amounts of electronic equipment can be subjected to considerable damage by lightning strikes. Other factors to consider are the temporary blinding effect at night, and the possible permanent error induced into the magnetic compass.

PRECIPITATION STATIC

When an aircraft flies through an area which contains clouds, precipitation, or a concentration of solid particles (dust, sand, ice, etc.), it accumulates a static electric charge. When this static electricity discharges onto a nearby surface or as a corona into the air, a noisy disturbance, called precipitation static, is picked up by radio receivers and interferes with radio reception, especially at lower frequencies. This is overcome by certain techniques of construction of aircraft, and by the use of wicks, antistatic antenna hardware, and VHF frequencies.

THUNDERSTORM SURFACE PHENOMENA

The rapid change in wind direction and speed immediately prior to a thunderstorm passage is a significant surface hazard associated with thunderstorm activity. The strong winds which accompany thunderstorm passages are the result of the horizontal spreading out of downdraft currents from within the storm as they approach the surface of the earth.

This spreading-out activity is depicted in Figure 187. Wind speeds at the leading edge of the thunderstorm are ordinarily far greater than those at the trailing edge. The initial wind surge observed at the surface is known as the first gust. The speed of the first gust is normally the highest recorded during the storm passage and may vary as much as 180° in direction from the surface wind direction which previously existed. The mass of cooled air spread out from downdrafts of neighboring thunderstorms (especially in squall lines) often becomes organized into a small, high-pressure area called a bubble-high or meso-high, which persists for some time as an entity that can sometimes be seen on the surface weather map. These highs may be a mechanism for controlling the direction in which new cells form.

THUNDERSTORM ALTIMETRY

During the passage of a thunderstorm, rapid and marked surface pressure variations generally occur. These variations usually occur in a particular sequence characterized by: (1) an abrupt fall in pressure as the storm approaches, (2) an abrupt rise in pressure associated with rain showers as the storm moves overhead (very often associated with the "first gust"), and (3) a gradual return to normal pressure as the storm moves on and the rain ceases. Such pressure changes may result in significant altitude errors on landing if the altimeter setting of a landing aircraft

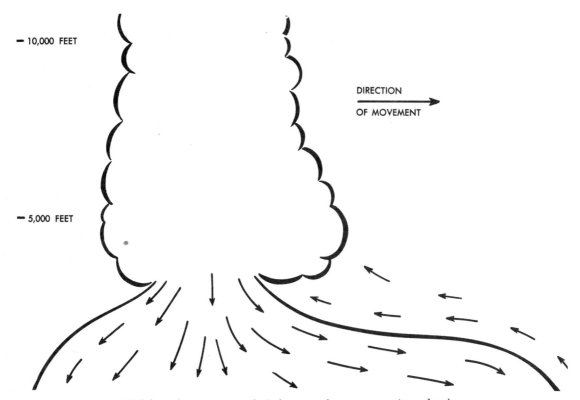

Fig. 187 Schematic presentation of winds near surface accompanying a thunderstorm.

is not corrected. (In general for the lower levels of the atmosphere, .10″ mercury is approximately equivalent to 100 feet of altitude.)

During a study of thunderstorms the maximum pressure rise and fall were converted to the equivalent altimeter error and tabulated. In 22% of the cases if a pilot landed during the maximum pressure using an altimeter setting given to him only a few minutes earlier, his altimeter would indicate that he was 60 feet or more below the true altitude.

Of greater concern to the pilot are pressure altitude readings which are too high. If a pilot used an altimeter setting given to him during the maximum pressure and landed after the pressure had fallen, on 26% of the days he would have found that his altimeter still read 60 feet or more above the true altitude after he was on the ground. On two occasions, the altimeter would have read over 140 feet above the true altitude when he landed.

TORNADOES

Tornadoes are violent circular whirlpools of air shaped like an inverted funnel from a cumulonimbus or associated cloud, and are commonly several hundreds of yards in diameter where they intersect the ground. The low pressure encountered in the center of the tornado and its high wind speeds are very destructive. Their paths over the ground are often only a few miles long, and they move at speeds of 25 to 50 knots. Although maximum wind speeds associated with tornadoes have never been precisely measured, property damage and other effects indicate that they probably exceed 300 mph. Tornadoes are bred from thunderstorms or other highly convective clouds in the form of twister (funnel clouds) which descend erratically to the earth and rotate in the counterclockwise (cyclonic) manner in the Northern Hemisphere.

Tornadoes are rare in most areas of the world, except Australia and the United States. In these areas a large number are reported each year. In the United States they occur most frequently in the Middle West and Southwest. They also occur frequently in the Southeastern States. Tornadoes are called waterspouts when they occur over water. However, all waterspouts are not tornadoes. Some waterspouts are formed by rapidly converging and convecting air, as are also the dust devils over land. These can rotate either clockwise or counterclockwise, and do not contain the destructive tornadic force.

Fig. 188 Waterspout at Biloxi, Mississippi.

Fig. 189 Tornado.

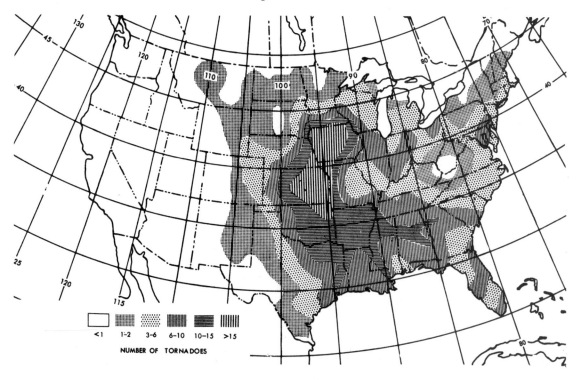

Fig. 190 Total number of tornadoes per 50-mile square reported in the period 1920-1949.

SOME GENERAL RULES

1. Whenever possible circumnavigate a storm. Always fly around isolated air mass thunderstorms.

Fig. 190A Circumnavigate.

2. When approaching a thunderstorm, analyze it before the surrounding clouds are encountered. Once you get into them, they may obscure important characteristics of the storm.

3. In coastal regions where thunderstorms prevail along the mountains, fly a few miles to seaward and avoid them.

4. Thunderstorms over islands may be thousands of feet higher than those over the open sea. Fly around them.

5. Cold front thunderstorms generally stretch too far to fly around. Remember that the storm front is a series of individual storms linked by intervening clouds. If you have to go through, fly between the storm centers or over the saddlebacks.

6. If you can't see blue sky beyond the storm and must go through, determine the direction the storm is taking and head in at a right angle.

7. Once you have headed into a storm, don't turn around on account of turbulence, rain or hail. If you do, you'll have to fly through the same conditions twice and you may get lost. Hold your original course.

8. In entering the front of a thunderstorm, you will encounter updrafts. Go in low, and if conditions permit, fly under the base of the storm.

9. In entering a storm area from the rear, you will experience downdrafts first. Go in high.

10. In flying under a storm, the higher the flight level, the rougher the trip. Fly about one-third of the distance from the ground to the base of the cloud if you can. But don't go underneath unless you can maintain contact flight.

11. Don't try to fly underneath a storm along mountain ranges unless there is a good ceiling and you can see peaks and ridges clearly. Over water you can usually count on being able to fly under any thunderstorms in daylight.

12. Never land at an airport when a thunderstorm is advancing toward the field. Shifting surface winds make it too hazardous. Wait until the storm center has passed and the winds have stopped shifting before you land.

13. When you expect to try high-level flight, get altitude before approaching the storm so that you are on top of the cloud shelf around the storm and can inspect the storm line before selecting your course.

14. The altitude necessary to fly around the tops and over the saddleback of a thunderstorm will vary with the seasons and the latitude in which you encounter the storm. In high latitudes, 12,000 to 15,000 feet is generally sufficient. In the tropics the tops of the saddlebacks may be above the ceiling of your aircraft and you may have to fly through them on instruments, a procedure recommended only for high performance aircraft. Over the open sea, 15,000 feet will usually clear the saddlebacks, except in the tropics.

15. Lightning is of little consequence when you are flying an all-metal, closed-cockpit plane. It acts as a perfect conductor. Switch on the cockpit lights and keep your eyes on the instrument panel so that the bright flashes won't blind you.

16. Avoid the freezing level in thunderstorms because of the dangers of icing and excessive turbulence.

In general, thunderstorm turbulence, blinding lightning, icing, heavy rainfall, and high winds present very definite hazards to flying. A basic knowledge of thunderstorm activity will help you cope with these hazards. However, the astute action is to avoid all thunderstorms if possible.

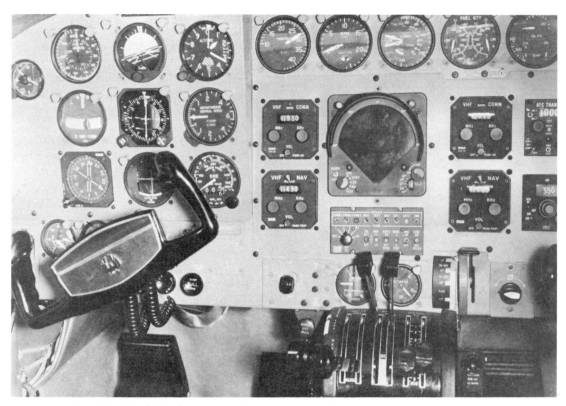

Fig. 191 Weather radar scope and controls as installed in a twin-engine Cessna 421.

AIRBORNE WEATHER RADAR

From a weather standpoint, hail and turbulence are the principal obstacles to safe and comfortable flight; yet, neither of these conditions is directly visible on radar. The radar only shows us the rainfall patterns with which these conditions are associated... *but that is all we need.*

The first truth to remember about weather radar is this: It can see water *best* in its liquid form. Not water *vapor*; not ice *crystals*; nor hail when small and perfectly dry. It *can* see rain, *wet* snow, *wet* hail, and *dry* hail when its diameter is about 8/10 of the radar wavelength or larger. (At X-band, this means that dry hail becomes visible to the radar at about one inch diameter.)

Now here are some truths about weather and flying: (1) Turbulence results when two air masses at different temperatures and/or pressures meet. (2) This meeting can form a thunderstorm. (3) The thunderstorm produces rain. (4) Your radar displays that rain, thus reveals the turbulence. (5) In the thunderstorm's *cumulus* stage, echoes will appear on the display and grow progressively larger and sharper. The antenna may be tilted up and down in small increments to maximize the echo pattern. (6) In the thunderstorm's *mature stage*, radar echoes will be sharp and clear; characteristic hail protuberances occur most frequently early in this stage. (7) In the thunderstorm's *dissipating stage* the rain area is largest and shows best with a slight down-

Fig. 192 Actual photo of a radar scope showing two rainfall areas. The farthest away at 80 to 100 miles, 15° right shows heavy rainfall in the center indicating a thunderstorm.

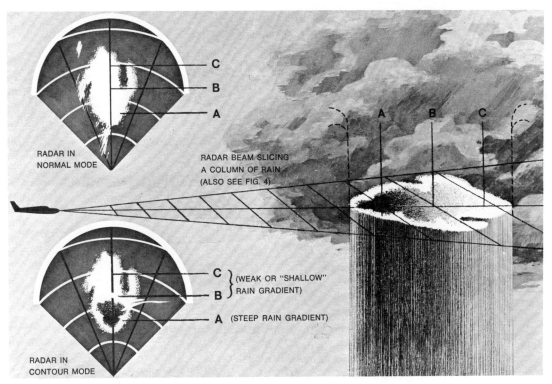

Fig. 193 A radar beam slicing a column of rain.

ward antenna tilt; contouring cells shrink and weaken while lighter-rainfall areas fade away.

Figure 193 helps to illustrate all of these points. We can use our radar to look inside the precipitation area to spot zones of present and developing turbulence. To identify these areas as turbulent, however, we must know a little more about meteorology.

The most important *single* fact is this: The areas of maximum turbulence occur where the *most abrupt* changes from light or *no* rain to heavy rain occur. The term applied to this change in rate is *"rain gradient."* The greater the change in rainfall rate, the steeper the rain gradient. The steeper the rain gradient, the greater the accompanying turbulence.

Rivaling this in importance is another fact: Storm "cells" are not static or stable conditions, but are in a constant state of change. While a single thunderstorm seldom lasts more than an hour, a squall line can contain many such storm cells developing and decaying over a much longer period. A single cell can start as a cumulus cloud only a mile in diameter, rising perhaps to 15,000 feet ... *but grow within ten minutes to five miles in diameter and a towering altitude of 60,000 feet or more.* Hence, weather radar should *not* be used to "take flash pictures of weather," but to keep

weather under continuing surveillance.

As these masses of warm, moist air are hurled upward to meet the colder air above, the moisture condenses and builds into raindrops heavy enough to fall downward through the updraft. When this precipitation is heavy enough, it can reverse the updraft. Between these downdrafts (shafts of rain), updrafts continue at tremendous velocities. It is not surprising, therefore, that the areas of maximum turbulence are near these interfaces between updraft and downdraft. *Keep these interesting facts in mind if you are ever tempted to crowd a rainshaft or to fly over an innocent-looking cumulus cloud!*

Since our primary purpose (Fig. 194) is to find a safe and comfortable route through the precipitation area, we should study the radar image of the squall line while closing in on the thunderstorm area. In this example, radar observation shows us that the rainfall is steadily *diminishing* on the left while it is very heavy in two mature cells (and increasing rapidly in a third cell) to the right. Obviously, our safest and most comfortable course lies to the left where the storm is decaying into a light rain. The growing cell on the right should be given a *wide* berth.

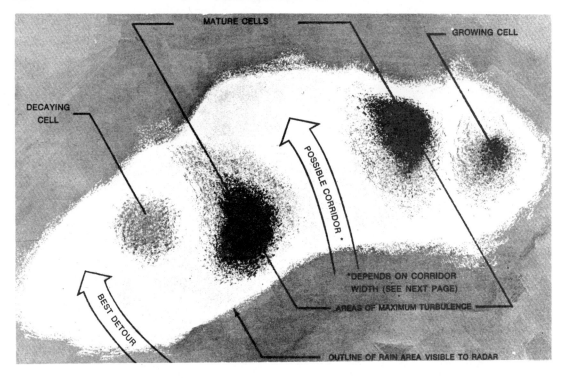

Fig. 194 Radar shows the safest and most comfortable paths through a thunderstorm area.

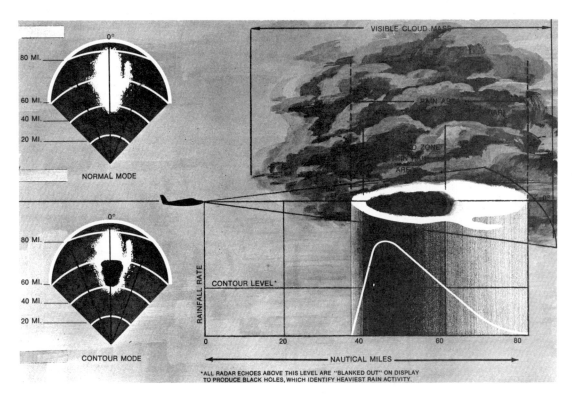

Fig. 195 "Contour mode" shows areas of heaviest rainfall.

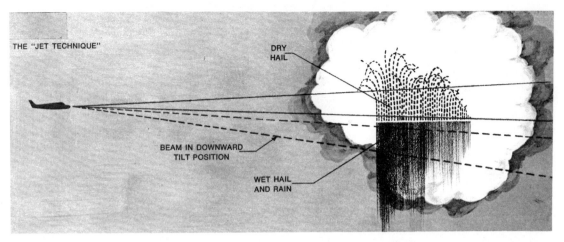

Fig. 196 *Since radar cannot "see" dry hail, a slight downward tilt of the antenna will reveal the rain resulting from the melting hail falling through lower, warmer air.*

After a certain rain density is reached, your radar will show an image of maximum brilliance. Radars with a CONTOUR mode of operation, however, can give us considerably more information on the storm. Figure 4 is a case in point. In this example, we have approached a typical thunderstorm with our radar in NORMAL mode (the *search* mode). A light rainfall at one edge of the cloud mass increases rapidly to a heavy downpour, then gradually diminishes into a light rain on the opposite side. On NORMAL or *search* mode (Fig. 195), however, our radar displays a large bright target extending from the 40-mile to 80-mile range marks . . . a 40-mile target which simply says, "Rain ahead." The radar returns contain much more useful information which can be brought out by switching to the CONTOUR mode.

When contouring, all signals above a preselected strength (i.e., signals from rainfall exceeding a pre-selected rate) are blocked. This rejection of signals by your radar's contouring circuitry makes certain that you see this area of maximum cell activity as a *black hole,* outlined by areas of lesser rainfall. Remembering what we learned from Figure 193, all we have to do now is to avoid that contoured area by a safe distance, especially on the steep rain gradient side. (NOTE: "Steep rain gradient" or "sharp shear zone" can be defined as a rain return with less than three miles between a contoured "hole" and a no-rainfall area.)

The safe distance for detouring, or the safe corridor width between cells, depends on several factors: Wind direction and speed; range

to target; radar efficiency; antenna size; radiated power; wave length; aircraft speed; available avoidance distance; condition of your radome; type of target; and the Outside Air Temperature (OAT). If the OAT is well below freezing, give *any* contoured area a wide berth (there could be dry hail around the cell, unseen by the radar). Even if the OAT is above freezing, watch closely for those "fuzzy" returns near cells; they *could* still be hail (see next section).

It is accepted practice, established by a major airline, to avoid all echoes showing a sharp shear zone by 5 nautical miles *when the OAT is 0°C or above.* When the OAT is *below* 0°C, avoid all echoes showing a sharp shear zone by 10 miles. Above 23,000 feet, avoid *all* echoes (contouring or not, with or without sharp shear zones) by 20 miles. Use these horizontal avoidance distances even when flying VFR *above* the weather with vertical clearances less than 5,000 feet. Any deviation from these basic practices depends on the *experienced* interpretation of weather radar returns.

Remember that the *search* mode for your radar is the NORMAL mode, and you should always return to this setting after locating cell activity on the CONTOUR mode. But remember also to switch back to CONTOUR mode regularly to check for cell activity ahead. Distant thunderstorms may not give strong enough radar returns to contour simply because they are too far away, and may resemble safe rain until checked for contouring at closer range.

As stated before, small dry hail will not return echoes on a radar which is designed for

Fig. 197 Radar patterns likely to contain hail.

weather avoidance. As it falls into warmer air, however, it begins to melt and form a thin surface layer of liquid which *will* give a return. A slight downward tilt of your antenna (toward the warmer air at lower altitude) may show rain coming from unseen dry hail that is *directly in your path!* (Fig. 196) When rain returns appear *below* your flight path, but not in your line of flight, you could be flying into hail. At *low* altitude operation, the reverse is sometimes true: Your radar may be scanning *below* a rapidly developing storm cell, from which the heavy rain droplets have not had time to fall to your flight level through the updrafts. Tilting the antenna up and down regularly will keep you posted on the total weather picture. If you are ever tempted to forget and leave your antenna pointed straight ahead throughout your flight, remember this warning.

Often, these hailstorms will take on familiar, easily identified patterns like those in Figure 197. Fingers or hooks are cyclonic winds radiating from the main body of a storm, and they *usually contain hail.* A "U" shaped pattern is also (frequently) a column of dry hail that returns no signal but is buried within a larger

area of rain which *does* return a strong signal. Scalloped edges on a pattern also strongly indicates the presence of dry hail bordering a rain area. Finally, weak or "fuzzy" protuberances are not *always* associated with hail, but should be watched closely; they can change rapidly.

The more we learn about radar, the more it seems that the pilot is an all-important part of the system. The proper use of controls is essential to gathering all pertinent weather data. The proper interpretation of that data (the displayed patterns) is equally important to safety and comfort.

This point is illustrated again in Figure 198. When flying at higher altitudes, a storm detected on the long-range setting may disappear from the display as it is approached. While it *may* have dissipated during your approach, don't count on it. It may be that you are directing the radiated energy from your antenna *above* the storm as you get closer. If this is the case, the weather will show up again when you tilt the antenna downward as little as one degree. Assuming that a storm has dissipated during the approach can be quite dangerous;

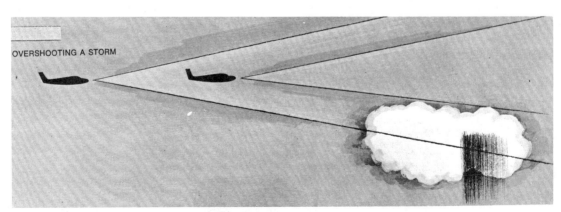

Fig. 198 Since there is also turbulence above a storm, downward tilt of the antenna may reveal a storm below the flight path.